D0324179

Information management in a contractor

A model of the flow of project data

Information management in a contractor

A model of the flow of project data

Norman Fisher and Shen Li Yin

Published by Thomas Telford Services Ltd, Thomas Telford House, 1 Heron Quay, London E14 4JD

Distributors for Thomas Telford Books are

USA: American Society of Civil Engineers, Publications Sales Department, 345 East 47th Street, New York NY10017-2398
Japan: Maruzen Co Ltd, Book Department, 3-10 Nihonbashi 2-chome, Chuo-ku, Tokyo 103
Australia: DA Books and Journals, 11 Station Street, Mitcham 2131, Victoria

First published 1992

A catalogue record for this book is available from the British Library

ISBN-13: 978-0-7277-1666-8

© Norman Fisher and Shen Li Yin, 1992

All rights, including translation reserved. Except for fair copying, no part of this publication may be reproduced, stored in a retrieval system or transmitted in any form or by any means, electronic, mechanical, photocopying or otherwise, without the prior written permission of the Publications Manager, Publications Division, Thomas Telford Services Ltd, Thomas Telford House, 1 Heron Quay, London E14 4JD.

The book is published on the understanding that the authors are solely responsible for the statements made and opinions expressed in it and that its publication does not necessarily imply that such statements and or opinions are or reflect the views or opinions of the publishers.

Preface

'Systems produce consistency and reliability, but people produce excellence ....'

This book describes a management information model developed as a result of a major research study. The authors hope that it will increase knowledge and stimulate further work. The book has been written for those who are concerned about the future direction of project management systems as used by contractors. It is also aimed at those who need a common denominator to compare their own company systems, to ensure that the 'best current practice' is used.

The model described is based on a systems analysis of observed systems and on an amalgamation of the systems of three leading contractors. It uses the 'structured data analysis' technique, a rigorous rule-driven systems analysis methodology. The authors believe that the development of this approach is an important step in the move towards computer-integrated construction and, in particular, the automation of construction project management information systems. Furthermore, the authors believe that this approach is the first step for the contractor towards harnessing the sort of productivity benefits of information technology already seen in manufacturing industry: 'lean manufacturing', 'flexible manufacturing' and 'knowledge based engineering'.

It is important to understand that this book is not intended as a 'best practice' manual for practitioners who are naturally concerned with more immediate short-term issues; nor is the model described more than simply a good standard against which to judge another system. The book is therefore aimed at corporate management consultants, corporate systems analysts, postgraduate students, forward thinking senior managers and also researchers, whether in a commercial research organization or a university.

The authors hope that their work will stimulate fruitful questioning of current methods and practices, and lead to more work that will add to and improve the model they describe. They are keen to stimulate international systems comparisons, both within and outside of the European Community.

Preface

Contents

1. Information technology systems and the management of construction sites 1
2. Mapping the flow of data 11
3. Defining the form of the data 54
4. Developing more robust systems 208
Appendix 1. The structured data analysis technique 218
Appendix 2. Reflections on the robustness of the model and on future directions 245
References 253
Glossary 257
Acknowledgements 259
About the authors 260

1. Information technology systems and the management of construction sites

Introduction

The first serious attempts at harnessing the benefits of information technology (IT) in the UK construction industry began in the late 1970s but progress during the 1980s has been superficial in many ways. So far IT has not had the sort of effect on construction that it has had on, for example, industrial engineering (refs 1 - 3). This is possibly because of the fragmented nature of the construction industry, the one-off nature of its product. By contrast, UK industrial engineering has been forced to adopt IT because of the following life or death factors.

(a) The success of the Japanese engineering/manufacturing industry in adopting systems and automation.
(b) The emergence in Japan of 'flexible manufacturing', where small batches of quite different products can follow each other down a manufacturing or assembly line, and 'simultaneous engineering', the buzz word for systems, where design, development, tooling, manufacture, assembly and quality are all considered at the same time. This technique is important because it reduces the key lead time between initiation of a project concept and selling in the market place. It ensures that the optimum blend of technologies is brought to bear on the complex process of creating, for example, a motor car.
(c) The benefits of the latest CAD software, such as knowledge-based engineering (KBE), that automates the simultaneous engineering strategy. This software system enables the building of knowledge-based models to incorporate not only shape and size, but also components, construction details, costing, production planning and user constraints. The objective behind such a model is to produce an optimized design and product (refs 4, 5) while reducing the design and construction time.

Two factors will greatly affect the UK construction industry over the next five to ten years: the globalization of construction and the move towards the site assembly of factory (pre-fabricated) components (ref. 6).

Rashid argues strongly that for the large construction company, there is little choice but to take globalization seriously (ref. 7). This means that UK contractors, whether they like it nor not, will increasingly be affected by, for example, US, Japanese, German and French contractors, consultants and component manufacturers. More components, design skills, technological systems and management will be traded across national borders in a cost effective way.

Moves towards site assembly will be accelerated by current activity in Japan. The pace in construction is likely to be set by Japanese contractors as it has been by Japanese industrial manufacturers, adapting IT for flexible manufacturing, simultaneous engineering and all aspects of KBE. For example, major Japanese manufacturing companies are moving into construction through the manufacturing of components. They are experimenting with the design and manufacture of buildings as systems with their advanced CAD expertise, and seem set to move fully into the construction business. Their industrial efficiency and capacity will ensure better value for money for clients. There is evidence to support this view in other countries. Researchers at Stanford University, for example, suggest that the US construction industry is set to follow the 'industrial engineering shift' (refs 8,9). It is most likely, however, that Japanese contractors will respond to the challenge first, probably by co-operating with the giant Japanese manufacturers.

A clear way forward for UK contractors is to follow the lead of UK and other European manufacturers, and develop a strategy of adopting IT in stages. This view is supported by the former National Economic Development office (NEDO) (ref. 5) which argued that the adoption of IT and electronic data interchange (EDI) will cut UK building costs by 15 to 25% (ref. 10). Two obvious ways for a contractor to start is either to systemize and automate the management process, or to automate the construction site. Significant work in each of these areas has begun in the UK (refs 11-14), Japan (refs 15,16), France (refs 17-19), Israel (ref. 20), and the US and Canada (refs 21, 22). Phenomenal advances are being made in the area of EDI. A new world of computer-driven data networks is springing up '...ranging in size from a tangle of cables joining a laboratory's workstations to a satellite system linking a multinational's laptops...'(ref. 23). These new networks provide magnificent opportunities for innovation.

A number of major research studies have demonstrated that the flow of data between the key members of a traditional construction project team is critically important for project success. One of these suggests that dealing with data occupies a large proportion of the time of those with important business and project managerial responsibilities (ref. 24).

This point has been thoroughly understood by the French contractor Bouygues. Bouygues has grown rapidly and successfully through acquisitions, and then by transferring to the recently-acquired company the Bouygues management system. This system incorporates a project management system, a computerized site data base and a management information system. Bouygues is a decentralized company both geographically and organizationally, and well known for its highly developed management expertise, personnel motivation and aggressive corporate culture. So the system is designed not to constrain managers, but rather to support them and bring consistency and reliability to the management operation. It is argued that this then 'frees up' the project engineers and managers to pursue 'excellence' (ref. 25).

The system that Bouygues has developed is both robust (well tested), and flexible (successfully used on a large number of different projects and in different countries). Bouygues claims that it has been an important factor in obtaining a competitive advantage. In the 'high risk' area of contracting, the consistency and reliability to be derived from automating, or systemizing, the management process would offer considerable commercial advantage benefits to a contractor primarily offering a management service (expertise). Consistency is also important for a contractor because of the problems caused by a fragmented industry. Research suggests that most problems of this type are at the interfaces between people, departments within an organization and between the various contractual parties (ref. 26).

Recent and significant advances in microprocessor capacity mean it is now possible to develop and operate sophisticated systems and models on small and inexpensive computers. There are two approaches to developing management systems in this way (ref. 12). The first starts from a clean sheet of paper and develops a logical, theoretically best system, and then by a series of field tests adapts it to operate successfully in a live situation. The advantage of this logical approach is that it takes new ideas, technology and freedom from old practices. However, its weakness is the inability fully to learn from experience, often gained over many years.

The second approach starts by modelling good current practice as observed, learning both strengths and weaknesses from it, why particular procedures, checks and so on have evolved, and then improves and automates the observed system. A criticism of this method is that bad industry practices may be inadvertently included. However, such dangers could be minimized by good systems analysis and an advisory expert panel.

Another criticism of this approach is that it is too restrictive and does not allow the radical benefits of new technology to be harnessed. There is certainly a danger of this happening if the researchers and analysts content themselves with an existing system, rather than using it as the distillation of current wisdom; as a starting point for developing more efficient systems. For this method to be effective, and to develop a holistic model of, for example, a contractor's project management system, any researcher will have to adopt a systems approach to the mapping of observed data (ref. 12). Note 1 considers why this is beneficial.

A holistic view of the data aspects of a contractor's management system for a major construction project will require a rule-driven research tool robust enough to cope with the partitioning, recording, mapping and analysis of very complex transmitted data. It will need to be in a form that allows repetition and challenge, and for important aspects of an organization to be compared with other similar organizations. Field tests have suggested that structured data analysis (SDA) meets very well the requirements outlined for mapping observed systems and for developing a holistic management system model (see note 2). Such a model is clearly important for practising contractors' project managers. There is direct evidence that a key skill of a successful project manager is the ability to take an overview of the system (or project) that he is trying to manage, and to understand how that system fits in holistic terms into both his parent organization and the project he is charged with managing (ref. 27) (see note 3).

A systems perspective

Behind the development of the model of the flow of data in a contracting company described in chapters 2 and 3 is the view that much can be gained by taking an overview and a systems view of how contractors obtain and manage construction work. Research work with thirteen major contractors suggested that their systems had developed in an incremental add-on manner and had not been looked at as a system, i.e. as a single entity, with the ensuing productivity benefits (see note 2). Further great benefit can be gained from comparing the commonality and the differences identified in the systems of different contractors, and also the type of work that they undertake. At a superficial level at least, most companies' systems appear to have little in common, in terms either of how they do business or how they manage construction. However, the research found more commonality where the systems were considered in terms of data, rather than in terms of what a managerial function is called, and who does it.

Perhaps the greatest achievement of this approach is the gathering and transfer of skill (knowledge and experience). By putting into the public domain good current practice in the form of a working system, this knowledge becomes a lowest common denominator in any discussion or study. There are obvious benefits for in-company training and development, as well as for university or professional education, and in a general sense for the medium and small-sized contractor. Witness the benefits to estimating since the publication, in 1965, of the CIOB's *Code of Estimating Practice* (ref. 28).

The model

Chapters 2 and 3 describe a general data flow model (GDFM) of a contractor's project management system. This is the combination of two or more individual companies' data flow models (DFMs) drawn from fieldwork into one common system, to represent good current practice. (For a detailed description of the technique see Appendix 1 and note 4.)

The GDFM is a map and specification of all the information that flows round a contracting company as part of a contractor's management system for a major construction project, and is viewed from the perspective of the information itself. Time-related, managerial control or human aspects are not included in any form. Only formal and informal data are shown, in a similar way to diagrams showing how blood flows around the human body. The data flow diagrams (DFDs) the maps of the observed flow of data in chapter 2, *start* with a context diagram showing how the subject data fits into the whole, and *end* with 'functional primitives', which are the lowest level of system partition possible; the smallest units into which data can meaningfully be split.

Chapter 3 introduces the data dictionary (DD). This is exactly what its name implies, a dictionary of terms used to describe and specify processes and flows of data. At the level of a functional primitive, the process notes describe what is happening to the data. So if the DFDs and the DD are combined, it is possible to look at a management system either on a superficial level or in detail. (Appendix 2 discusses the possible shortcomings of this approach.)

How to use the model

As a first step in understanding what the model shows, it is important to be familiar with the contents of Appendix 1. At the beginning of Chapters 2 and 3 is a description of each of the components of the model in terms of what it is, what it shows and how it can be used (see note 5.)

By following how the model is broken down (see the numbering of DFDs and processes in chapter 2) a number of key areas of interest such as pre-tender, pre-construction and construction procedures can be identified. It is then possible to cross-reference these with descriptions of processes, files and data flows in chapter 3. The information flows for key contracting processes can then be examined in detail. If a similar exercise is undertaken for all or part of the corresponding system in the reader's own company, direct comparisons can be made.

It is likely that most systems compared in this way will have major areas of commonality, but of equal and perhaps more interest, they will have some areas of difference. A GDFM is likely to have, on balance, more highly developed sub-systems than an individual company DFM, but not necessarily. Certainly the authors believe that the GDFM as presented in this book could be considerably improved by further systems comparison. It is quite a simple task to 'cut' out the inferior sub-system in the GDFM and 'paste' in a better, more developed one when it has been identified.

The first step in this improvement task is to identify common data flows between the DFM and the GDFM. This is a relatively simple task, as it involves identifying common terminators on the two systems. They are best visible on context diagrams (see Appendix 1) or as high a level of diagram as possible, i.e. the relevant sources and sinks (see Glossary). Once common terminators or data flows have been identified it is easy to connect the new sub-system into the bigger existing system at those common interface points.

The common data flows, although transmitting and receiving identical data in each case, can use different data element names in each DFM and GDFM. This needs to be allowed for by the detailed examination of the data in terms of form and content to establish that it is identical. It is possible to combine two or more GDFMs by linking them at their interface points, provided that the data flowing across the interface are compatible. Once integration has been achieved at the DFD level the two DDs are combined and adjusted accordingly.

A successful way of using the GDFM model in this book would be to set up within an interested company a task force to deal with a particular issue, cost/value reconciliation, and compare the data flows and processes of the relevant parts of the model with current practice in the firm. The next step would be to use the expert opinion in the task force to seek out the relevant merits of the two methods and decide on any changes or improvements that the company may wish to make.

Notes

1.There is considerable evidence that the application of systems ideas to an existing discipline can develop the level of understanding within it dramatically. This is particularly so in the unrestricted sciences, due largely to the complexity of the subject matter. Where systems ideas have been applied to unrestricted sciences such as biology or geography, to specific industries such as shipbuilding, or to companics and industrial social systems generally, such as coal-mining or the motor industry, the level of understanding has increased dramatically. The Cambridge geographers' rewriting of their subject is a case in point (refs 29-33). It is clear therefore that a systems approach to construction is likely to be beneficial.

2. A ten-year research programme at the University of Reading included more that 145 studies at the various stages of the construction procurement process. These included the feasibility study, evaluation of client needs, selection of design consultants, construction phase and post-construction phase. Results suggest that SDA meets well the requirements of a research tool to assist the development of systems models - the foundation for automation of a contractor's project management functions.

3. The management of construction is a complex business. Each construction project is a one-off in terms of client, location requirements and architectural design philosophy. Construction is a fragmented industry, and too often an adversarial culture exists at all levels. Furthermore the perception by city institutions of the UK construction industry as a place to deposit investor's funds is poor: low skill; low commitment to training other than at the most basic craft level; low commitment to research and development; high risk for investment and poor product performance and client satisfaction.

An unpublished study part of the fieldwork for the Image section of the 2001 series Task Force report (ref. 6) shows this. To warrant investment and as compensation for this perceived high exposure risk, companies are expected to achieve and maintain a high 'earnings per share' ratio. Consequently a very high proportion of profits in this sector is distributed to shareholders, often at the cost of training and research and development.

Possibly because of this scenario, little attention has been paid to the contractor's project management process as a whole. Research at the University of Reading has considered aspects of a contractor's project management process in detail, particularly tendering, scheduling and resource planning. Most of the recent studies have been stimulated by information technology advances. With interest in computer-integrated construction, construction project management systems, construction engineering systems, construction robotics and production automation now worldwide, there are unprecedented opportunities to develop more efficient methods of data transfer within a construction company and in so doing to achieve greater understanding of the issues involved.

Medawar, Popper and Kuhn (refs 34-37) give in their different ways a clear description of current scientific method, and suggest that if 'good science' is to develop, good theoretical 'ideal' models that develop logical 'best' or 'ideal' solutions will need to be balanced with experimental empirical work, that draws principles from an analysis of the 'best of current practice'. Medawar discusses in simple terms the principles of good scientific investigation; Popper concerns himself with the question of challenge, public domain knowledge and refutation (refs 34,36); Kuhn is concerned with the growth of knowledge by the building and challenging of current paradigms. The body of knowledge is developed in two mature restricted sciences, namely physics and astronomy. Both have a theoretical school and an empirical or experimental school. Often the theoretician may be many years ahead of his empirical colleague in terms of ideas and provisional explanations about aspects of his discipline. He may suggest, for example, the existence of some phenomenon 20 years before his experimental colleague can support this idea or explanation by empirical data. But the empirical scientist may demonstrate by experimentation that a long held, deeply cherished theory is no longer defendable.

From the 'tension' that sometimes exists between the two approaches, scientific opinion suggests, breakthroughs in understanding can and usually do occur. This book demonstrates a process for developing experimental work, powerful empirical models of data flows which will eventually represent better than the best of current practice, and combined with theoretical work will contribute to better systems understanding and design.

4. As the principal aim was to map, specify and record the monitoring and control procedures, the interviews of company staff were conducted on a top-down basis. Structured interviews were prepared, with suitable meeting agenda circulated to the relevant staff before the interviews began. Lists of other required information were prepared. These included such items as standard meeting prompts, standard agenda forms, company policy and procedure manuals.

At the fieldwork stage an initial interview was conducted with a senior director (usually the managing director) and also the person nominated by the company as the 'link person'. Representative staff were identified at each level and interviewed in the following order whenever possible: managing director, construction director, surveying director, buyer, accountant, estimator, plant manager, area construction manager, area surveyor, two project managers and two site agents (or senior foremen).

A total of 14 people was considered the most that could be interviewed in each company without straining the goodwill of the subject contractor, and within the research time constraints; and as sufficient to establish a clear picture of that company.

The interviews used well-established structured interview techniques, so as to minimize any interviewer bias. All interviews were conducted by two people and recorded on audio cassettes. On the basis of the recorded information and the written data gathered, levelled DFDs were constructed with context diagrams and process notes for all functional primitives (for a description of these terms and the SDA technique, see Glossary and Appendix 1).

Checks, known as structured walkthroughs (ref. 38), were conducted with the company link person, by telephone and face-to-face contact, to test the conceptual accuracy of the DFDs as well as the descriptive accuracy of both the DDs and the process notes.

The identified system was written up in draft form and a final structured walkthrough was conducted with each company, usually with the staff that had been interviewed. Once the fieldwork had been completed, the analysis of the contractor's management system for a major construction project could begin with the production of a common DD for the common flows, processes, files and terminators. Once established, the data flows and processes in the common DD were each given a new common name, usually drawn from the names in the system with the greatest common number.

With areas of commonality established, areas of divergence and systems bias were clearly visible and incorporated into the common system. Often this meant redrawing the DFDs. If two sub-systems were found to be in competition, then the more sophisticated in terms of flows (i.e. the one that had the need for more partitioning and containing more functional primitives) was selected.

5. The GDFM of a contractor's project management system was constructed by first identifying the common DD data elements between the three companies, and for sub-systems where commonality did not exist, selecting and adding to the common DD data elements the most sophisticated sub-systems of the three companies. Where a particular sub-system occurred in only one or two of the three companies, it (or the more sophisticated sub-system of the two) was added to the DD. Once the GDFM's DD had been compiled, the GDFM's DFDs were constructed by drawing on the three DFM's DFDs. When the common DFDs had been constructed, the draft GDFM was complete. The draft GDFM was at this stage checked for system balance, logic errors and inconsistency (ref. 26). Systems analysis workbench software is commercially available from at least two software houses and typically runs on a large and fast IBM PC compatible. The software used in this study was the *Yourdon Technique* software package. Its particular value lies in its ability to speed up the DFM construction process, and validate the model in terms of logical errors, balancing and consistency between DD and DFDs.

2. Mapping the flow of data

Introduction

There are many ways in which a managerial information system, or aspects of it, can be represented graphically. For consistency any methodology must be rule driven. Network analysis methods, representing time and logic, and workstudy, representing 'activity' and its classifications, widely used. A novel technique is linear responsibility analysis, which maps lines of managerial responsibility.

Walker and Hughes used this technique to develop retrospective case studies of project organization structures of private industrial projects (ref. 39). They looked at organizational systems rather data systems. They looked at people, the key players making strategic decisions, and at patterns, of roles, responsibilities and relationships, that combine to form an organizational structure for a project. They suggest that a changing project environment requires project organization structures to be responsive and dynamic. In essence, Walker and Hughes are analysing and describing the important systems used for the procurement of buildings – organizational systems.

By contrast the systems described in this book are data systems, viewed from the perspective of data flowing through and around a management information system. No account is taken of people, time, managerial control or resources, although the data contained within the system will obviously be essential for managers if they are to perform effectively.

What is a data flow diagram?

Chapters 3 and 4 describe a map of data flow (the DFDs) and the data that flow (the DD). This differs from the technique of Walker and Hughes in that it describes the system that construction project managers typically spend 70% of their time dealing with—data. (The generating, managing, sending, collecting and analysis of data is a major activity for almost all managers. This point is explained and demonstrated fully in Appendix 1.) It describes the systems which will produce the 'consistency and reliability'; and liberate the managers to produce the excellence talked of. Many analogies can be seen at key managerial points in other high risk industries, for example, commercial aviation and hi-tech military warfare. The modern airline pilot flying a new Boeing 747 - 400 series aircraft is really a systems manager and systems operator. By systemizing the routine and developing sub-system routines for every imaginable scenario, the human factor has been reduced, and consistency and reliability have consequently been increased. The pilot is therefore freed up to monitor, to look ahead and to have a better overall grasp, free of a mass of details, of the flying and minute by minute status of a commercial airliner with 350 or more passengers and crew.

A DFD can be summarized as a network or map of related functions showing all interfaces between components. It is also a partitioning of a system and its component parts.

The data model illustrated in this chapter is graphically portraying the flow of data (lines and arrows) between points where those data are altered or used in some way (circles). For example, in Dfd : 1 an arrow leaves bubble number 1.3 (planning procedures) and goes to bubble 1.6 (tender finalization). The data being transmitted are called resource data for project strategy. They will include not only precise data about 'preliminaries' requirements, but also a site layout plan, a method statement, a pre-tender programme and programme alternatives.

A practical example

The contents of the data element resource data for project strategy and how they are precisely recorded as part of this model, is described in Appendix 1. These diagrams show the flow of data around a contractor's organization, much like the system of veins and arteries that allow blood to flow around the human body. The graphical/logical technique which records it is an adaption of a computer systems analysis technique, known as structured data analysis. (Appendix 1 explains this technique and should be read and/or understood before proceeding.)

The model in this chapter

The diagrams show the flow of all data, formal information (e.g. forms, memoranda) and informal information (e.g. conversations, telephone calls), as it happens. The analyst that constructs such a diagram follows each piece of data in turn around an organization. It is purely the flow of data that is recorded, and not time-related events, managerial decisions or responsibility. The approach does, however, look at data as systems: for example, the data that flow as part of a cost control system or an estimating system.

All DFDs included in this chapter are numbered in such a way as to enable the partitioning logic to be followed. In several places DFDs have been split, as they are too large to fit clearly on to one page. For example, DFD #3 consists of four sheets, but only the first page is numbered, but the next three pages are all integral to DFD #3, entitled construction procedure.

The diagrams start with the most holistic view (see context diagram and Dfd # 0).

As the data are broken down, so more detail is shown of a particular contractor's managerial activity. For example, in Dfd # 0 process 3, this process is broken down or partitioned into eight sub-processes.

(a) 3.1 initiating project controls
(b) 3.2 site management
(c) 3.3 external valuation procedures
(d) 3.4 cost value comparison procedures
(e) 3.5 resource management
(f) 3.6 project information procedures
(g) 3.7 external progress meetings
(h) 3.8 internal site review meeting.

Some of these items are complex activities (systems) in themselves, and need to be further partitioned to allow more detailed analysis. For example, (b) can be broken down into seven further sub-processes (see Dfd # 3.2)

(a) 3.2.1 determine site strategy
(b) 3.2.2 site scheduling
(c) 3.2.3 forecasting resource requirements
(d) 3.2.4 co-ordinate site team
(e) 3.2.5 site personnel management
(f) 3.2.6 site performance monitoring
(g) 3.2.7 site information management.

At the information level of (g), the model is showing specific, detailed information. For example, a data flow from (b) to (c) is called resource requirement, and would deal with such issues as material requirements, sub-contractor requirements, plant requirements and information requirements, all in some detail.

How can the data flow model be used?

This model amalgamates the best features of the construction project management information systems of three major contractors. It is thus a reference point, or benchmark, against which a company can compare its overall system (or aspects of it), and perhaps improve its current system. It may be that aspects, or even the whole, of a company's existing system will be considered technically superior or preferable. There are several advantages to the exercise.

(a) The system will probably be improved.
(b) The operational managers will have spent time comparing their system with that of competitor companies.
(c) The operational managers will have greater confidence in their system and better understanding of it.
(d) The operational managers will have benefited from looking at their system in a holistic way. Research has shown that most systems have developed in an incremental way, perhaps best described as a succession of 'bolting on' operations (See Appendix 1). From a holistic view, it is easier to see logic errors, gaps and considerable built-in redundancy, caused by, for example, overlaps (people being given unnecessary information, or the generation of data that are little or never used).
(e) The development of systems thinking in contracting companies, by operations managers as well as senior staff, is vital if the companies are to remain competitive. Productivity improvements due to IT advances, attained by other UK, European and Japanese contractors will have to be matched. Once a working understanding of the SDA technique has been achieved, it will be possible to compare the company's existing system (its DFM), with the GDFM illustrated in this book (see Glossary).

There are two methods of comparison which experience has shown will produce good practical results.

The first uses existing company forms, current procedure manuals and the systems knowledge of an expert task force, for example on cost value reconciliation, and compares item by item, procedure by procedure, the way the company performs the operation against the logic displayed in the model. Clearly the model does not have the low level detail that would be available to an expert task force. The exercise should attempt to look at for example: who information is sent to, how much it is used and how often; whether the data being transmitted are in the most helpful format; instances of overlap, i.e. the same data being created and/or sent twice. Perhaps the same information could be sent to a purchasing department on a site daily report by a section engineer, as is sent by a site surveyor as part of his 'sub-contract management' system. Data bases (files) also create overlap. On a large project, six or more people may be keeping broadly the same data file, but each will jealously guard his own information source. A single site (computerized) data base, with access for all relevant people on one or more terminals or laptop computers, can be more efficient.

The first comparison approach is probably faster than the second approach and easier to undertake, but is not so rigorous and therefore less accurate. The second comparison method involves undertaking a full SDA of a company's own system to produce a DFM and then to compare subsystem with sub-system, process with process, with the GDFM. Much of the detailed comparison will be conducted in the same way as in the first method. The second approach, however, is likely to produce much more accurate results because of the more rigorous way in which the system of the subject company is analysed and mapped. The actual conducting of the systems analysis study needed to prepare the DFM will ensure that each component will be looked at in detail, on its own merits, but also as part of, and how it fits into an overall (holistic) system. However, this approach takes much longer than the first and requires some, although limited, systems analysis skills.

The data dictionary

Each element in a DFD, whether process, data flow, file, source or sink, is named and listed in the DD. When a data element exists at a point where a sub-system cannot be further decomposed (broken down), it is called a functional primitive. All functional primitive entries in a DD contain a mini-specification (mini-spec), i.e. a textual description of what the data element contains, what it does, what is done with it and, most important, the format or structure of the data.

Once the 'logic' of the flow of data within a particular aspect of the GDFM model has been understood, it is important to cross-reference the logic of the DFDs with the relevant descriptions in the DD. To use an analogy, the DFD is the skeleton on which the flesh of the DD is hung.

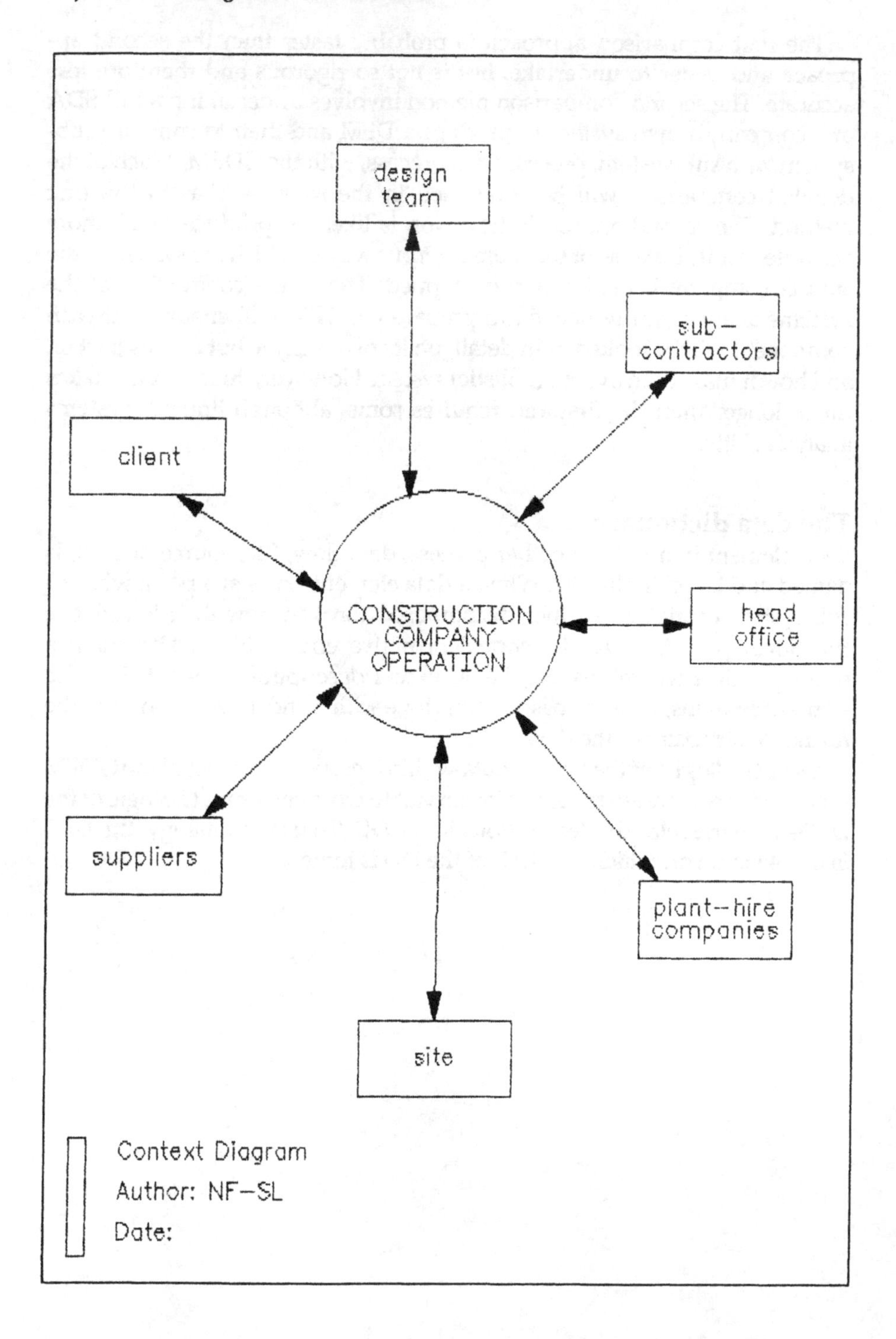

design team
sub-contractors
client
CONSTRUCTION COMPANY OPERATION
head office
suppliers
plant-hire companies
site
Context Diagram
Author: NF-SL
Date:

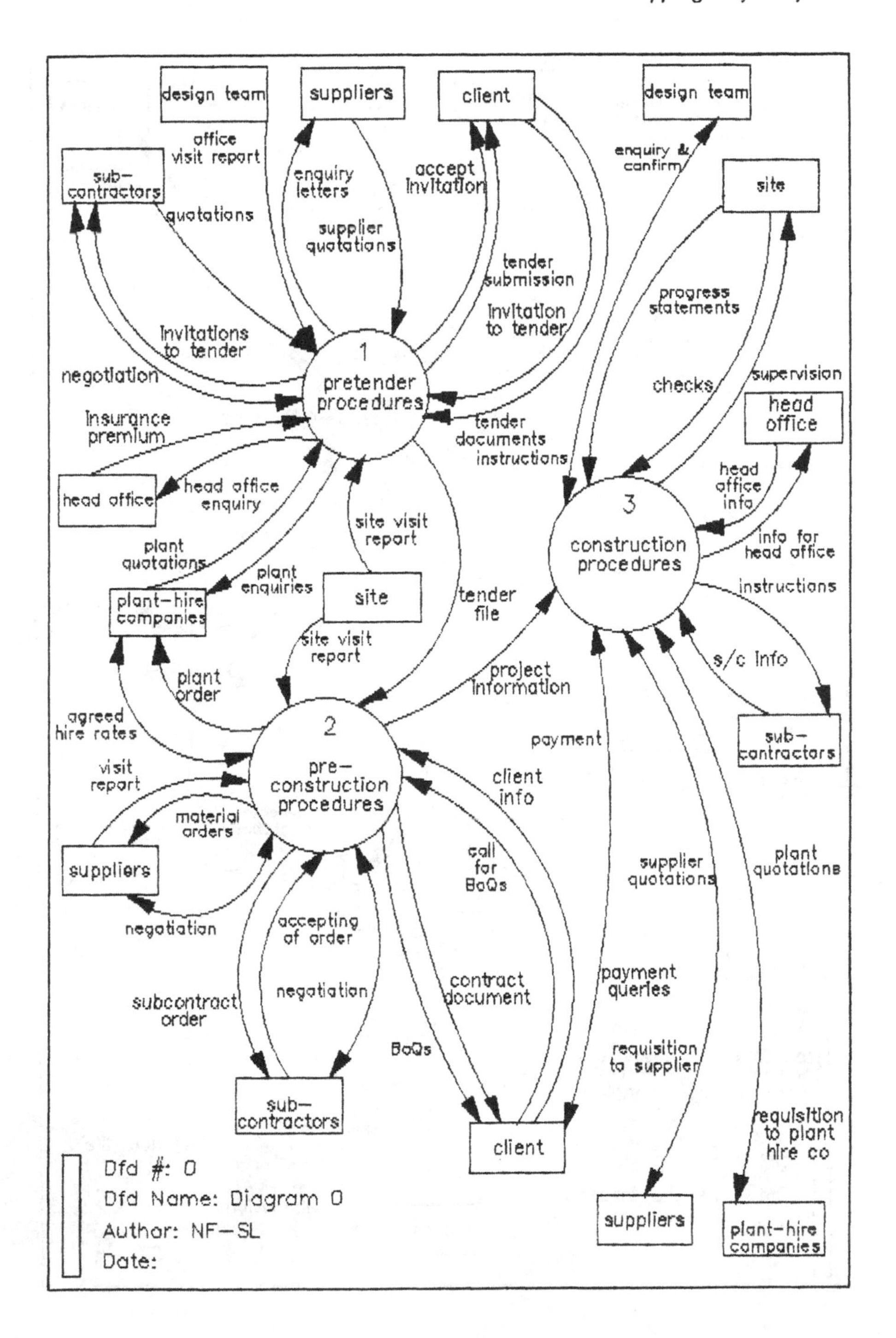
design team
suppliers
client
design team
sub-contractors
office visit report
enquiry letters
accept invitation
enquiry & confirm
site
quotations
supplier quotations
tender submission
progress statements
invitation to tender
invitations to tender
1
pretender procedures
negotiation
checks
supervision
head office
insurance premium
tender documents
instructions
head office
head office enquiry
head office info
3
construction procedures
site visit report
plant quotations
info for head office
plant enquiries
plant-hire companies
site
tender file
instructions
site visit report
s/c info
project information
plant order
agreed hire rates
2
pre-construction procedures
payment
sub-contractors
visit report
client info
material orders
suppliers
call for BoQs
supplier quotations
plant quotations
negotiation
accepting of order
contract document
payment queries
subcontract order
negotiation
BoQs
requisition to supplier
sub-contractors
client
requisition to plant hire co
Dfd #: 0
Dfd Name: Diagram 0
Author: NF-SL
Date:
suppliers
plant-hire companies

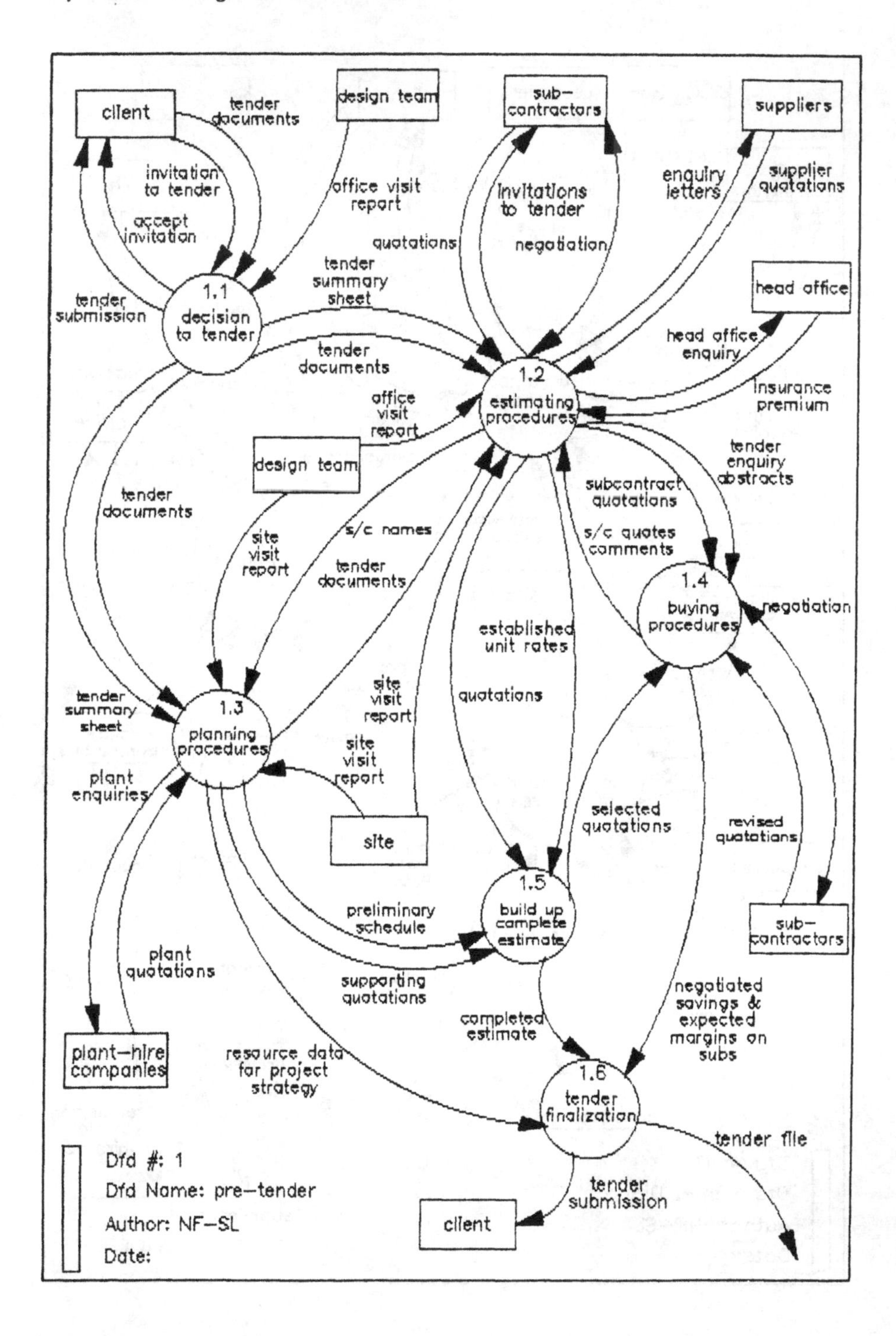
client
tender documents
design team
sub-contractors
suppliers
invitation to tender
accept invitation
office visit report
Invitations to tender
quotations
negotiation
enquiry letters
supplier quotations
tender submission
1.1 decision to tender
tender summary sheet
tender documents
head office
head office enquiry
insurance premium
1.2 estimating procedures
office visit report
design team
tender enquiry abstracts
subcontract quotations
s/c quotes comments
tender documents
site visit report
s/c names
tender documents
1.4 buying procedures
negotiation
established unit rates
tender summary sheet
1.3 planning procedures
site visit report
site visit report
quotations
plant enquiries
site
selected quotations
revised quotations
1.5 build up complete estimate
preliminary schedule
sub-contractors
plant quotations
supporting quotations
negotiated savings & expected margins on subs
completed estimate
plant-hire companies
resource data for project strategy
1.6 tender finalization
tender file
Dfd #: 1
Dfd Name: pre-tender
Author: NF-SL
Date:
tender submission
client

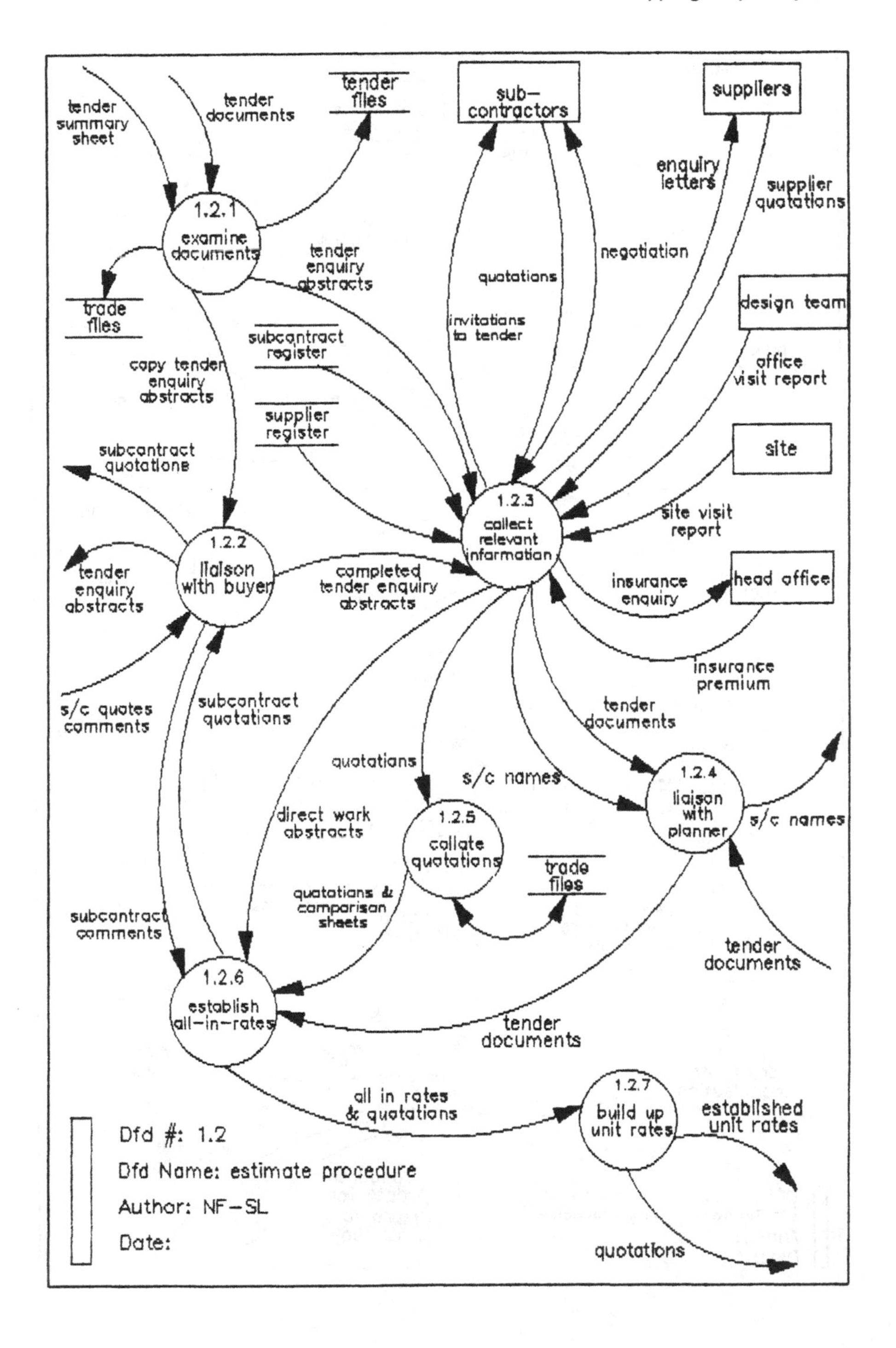

tender summary sheet
tender documents
tender files
sub-contractors
suppliers
enquiry letters
supplier quotations
1.2.1
examine documents
negotiation
tender enquiry abstracts
quotations
trade files
design team
invitations to tender
subcontract register
office visit report
copy tender enquiry abstracts
supplier register
site
subcontract quotations
1.2.3
collect relevant information
site visit report
1.2.2
liaison with buyer
tender enquiry abstracts
completed tender enquiry abstracts
insurance enquiry
head office
insurance premium
s/c quotes comments
subcontract quotations
tender documents
quotations
s/c names
1.2.4
liaison with planner
s/c names
direct work abstracts
1.2.5
collate quotations
trade files
quotations & comparison sheets
subcontract comments
tender documents
1.2.6
establish all-in-rates
tender documents
all in rates & quotations
1.2.7
build up unit rates
established unit rates
Dfd #: 1.2
Dfd Name: estimate procedure
Author: NF-SL
Date:
quotations

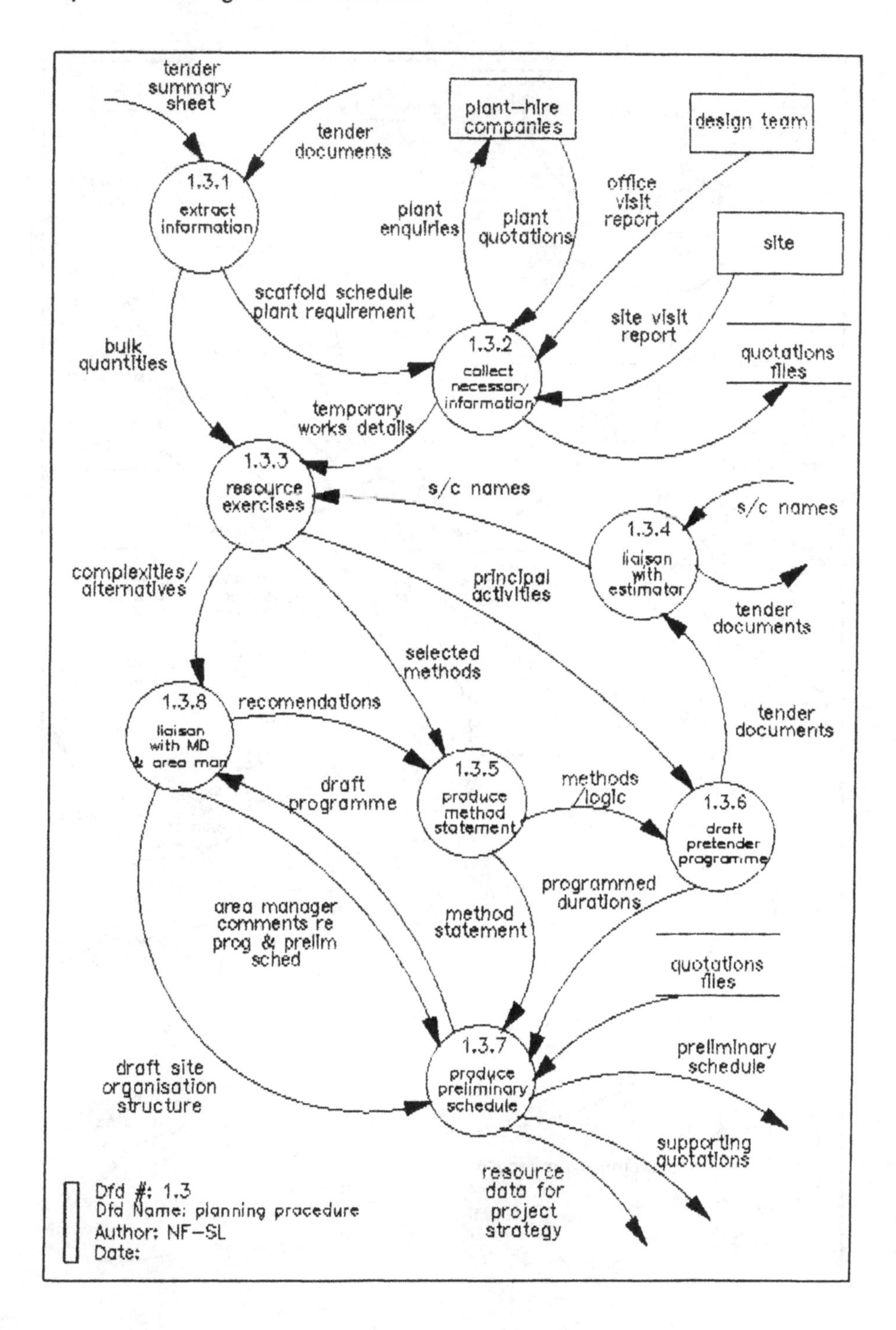

tender summary sheet
tender documents
plant–hire companies
design team
1.3.1 extract information
plant enquiries
plant quotations
office visit report
site
scaffold schedule plant requirement
site visit report
bulk quantities
1.3.2 collect necessary information
quotations files
temporary works details
1.3.3 resource exercises
s/c names
s/c names
1.3.4 liaison with estimator
complexities/ alternatives
principal activities
tender documents
selected methods
1.3.8 liaison with MD & area man
recomendations
tender documents
draft programme
1.3.5 produce method statement
methods /logic
1.3.6 draft pretender programme
programmed durations
area manager comments re prog & prelim sched
method statement
quotations files
preliminary schedule
draft site organisation structure
1.3.7 produce preliminary schedule
supporting quotations
resource data for project strategy
Dfd #: 1.3
Dfd Name: planning procedure
Author: NF–SL
Date:

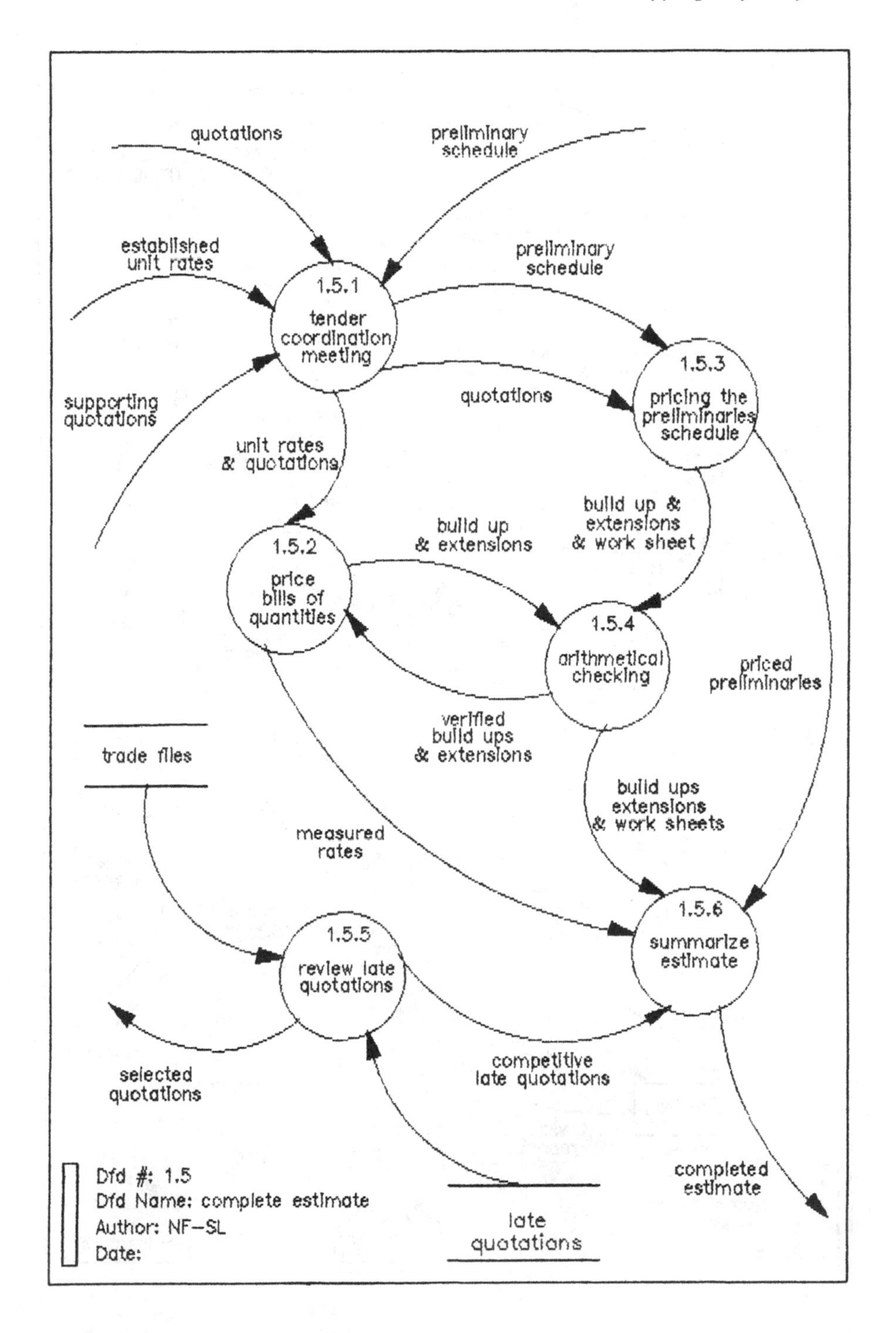
quotations
preliminary schedule
established unit rates
1.5.1 tender coordination meeting
preliminary schedule
supporting quotations
quotations
1.5.3 pricing the preliminaries schedule
unit rates & quotations
build up & extensions & work sheet
build up & extensions
1.5.2 price bills of quantities
1.5.4 arithmetical checking
priced preliminaries
trade files
verified build ups & extensions
build ups extensions & work sheets
measured rates
1.5.6 summarize estimate
1.5.5 review late quotations
selected quotations
competitive late quotations
completed estimate
late quotations
Dfd #: 1.5
Dfd Name: complete estimate
Author: NF–SL
Date:

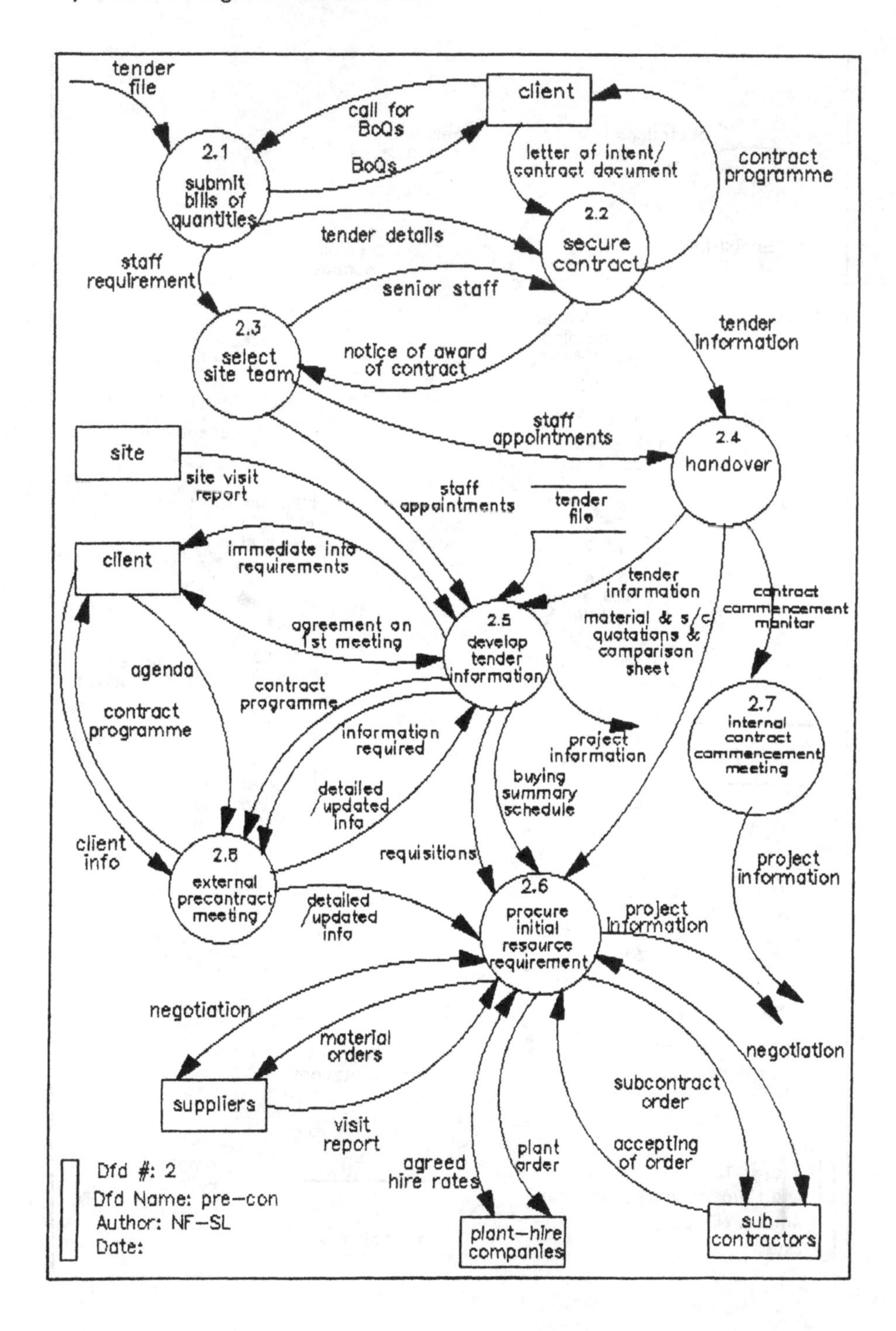

tender file
client
call for BoQs
BoQs
2.1 submit bills of quantities
letter of intent/ contract document
contract programme
2.2 secure contract
tender details
staff requirement
senior staff
2.3 select site team
notice of award of contract
tender information
staff appointments
2.4 handover
site
site visit report
staff appointments
tender file
client
immediate info requirements
tender information
contract commencement monitor
agreement on 1st meeting
2.5 develop tender information
material & s/c quotations & comparison sheet
agenda
contract programme
contract programme
information required
project information
2.7 internal contract commencement meeting
detailed /updated info
buying summary schedule
client info
requisitions
project information
2.8 external precontract meeting
detailed /updated info
2.6 procure initial resource requirement
project information
negotiation
material orders
negotiation
suppliers
subcontract order
visit report
plant order
accepting of order
agreed hire rates
Dfd #: 2
Dfd Name: pre-con
Author: NF-SL
Date:
plant-hire companies
sub-contractors

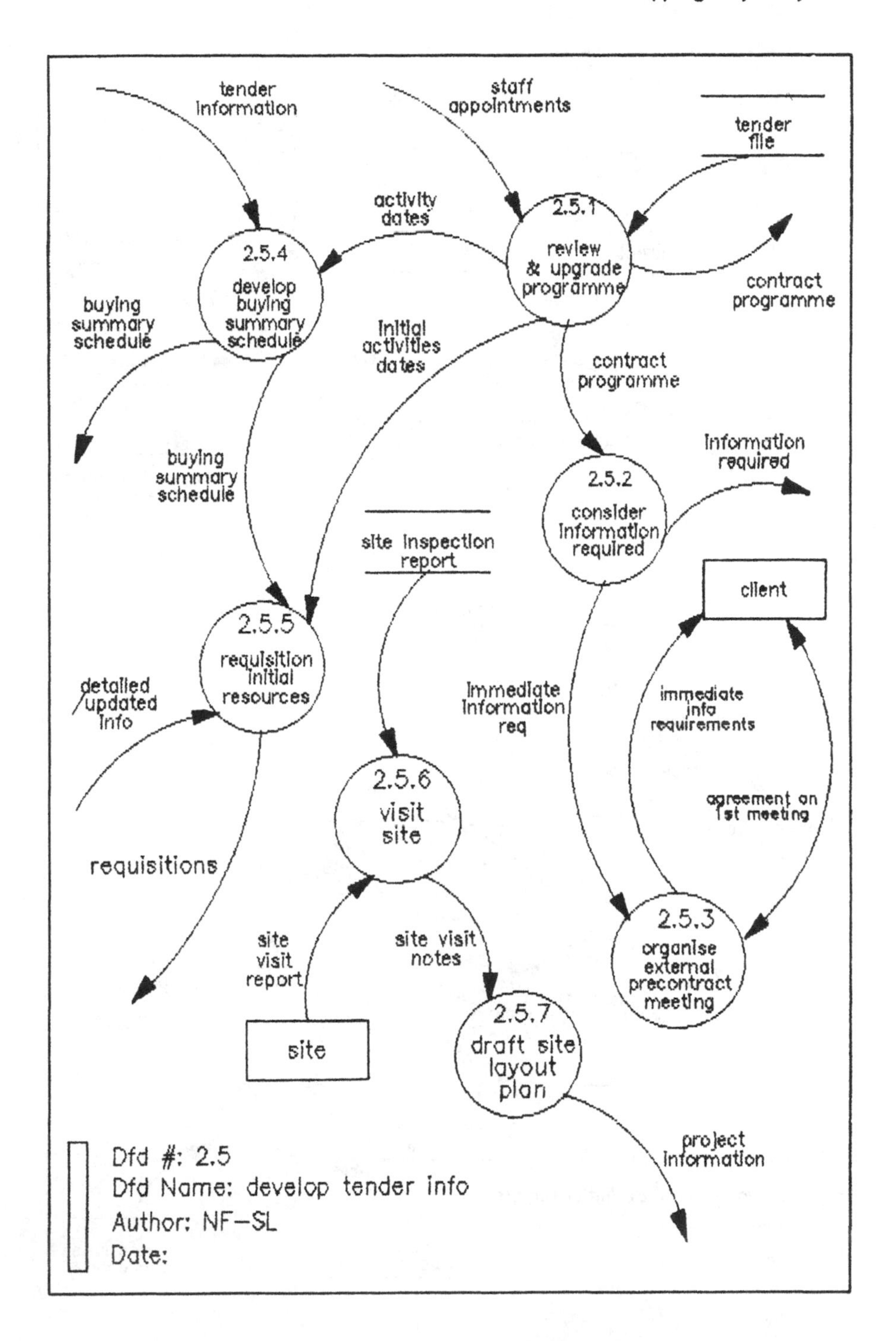
tender information
staff appointments
tender file
activity dates
2.5.1 review & upgrade programme
contract programme
2.5.4 develop buying summary schedule
buying summary schedule
initial activities dates
contract programme
information required
buying summary schedule
2.5.2 consider information required
site inspection report
client
2.5.5 requisition initial resources
detailed /updated info
immediate information req
immediate info requirements
agreement on 1st meeting
2.5.6 visit site
requisitions
2.5.3 organise external precontract meeting
site visit report
site visit notes
site
2.5.7 draft site layout plan
project information
Dfd #: 2.5
Dfd Name: develop tender info
Author: NF-SL
Date:

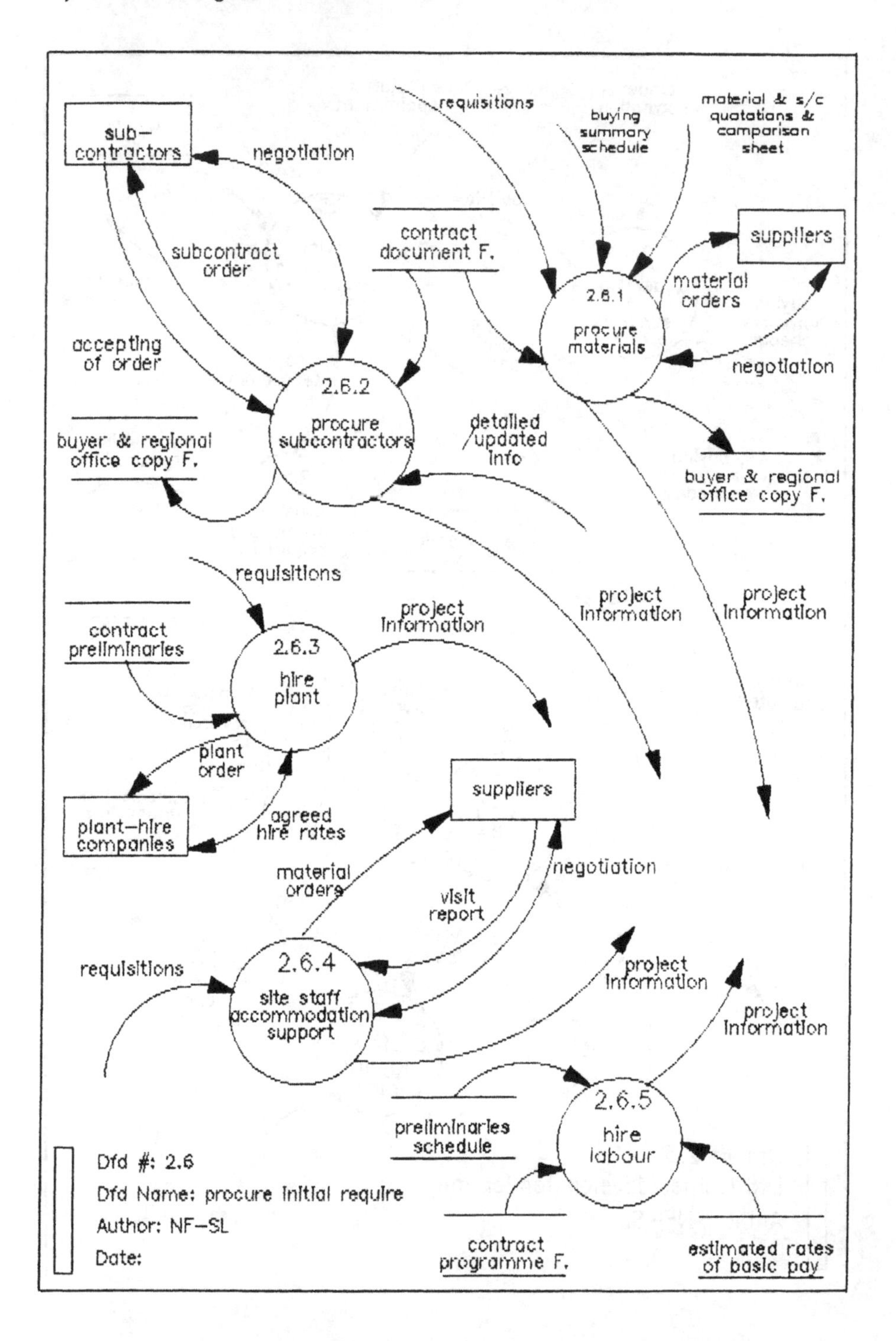

sub-contractors
negotiation
requisitions
buying summary schedule
material & s/c quotations & comparison sheet
contract document F.
suppliers
subcontract order
material orders
2.6.1
procure materials
accepting of order
negotiation
2.6.2
procure subcontractors
detailed /updated info
buyer & regional office copy F.
buyer & regional office copy F.
requisitions
project information
project information
contract preliminaries
2.6.3
hire plant
project information
plant order
suppliers
plant-hire companies
agreed hire rates
negotiation
material orders
visit report
requisitions
2.6.4
site staff accommodation support
project information
project information
2.6.5
hire labour
preliminaries schedule
Dfd #: 2.6
Dfd Name: procure initial require
Author: NF-SL
Date:
contract programme F.
estimated rates of basic pay

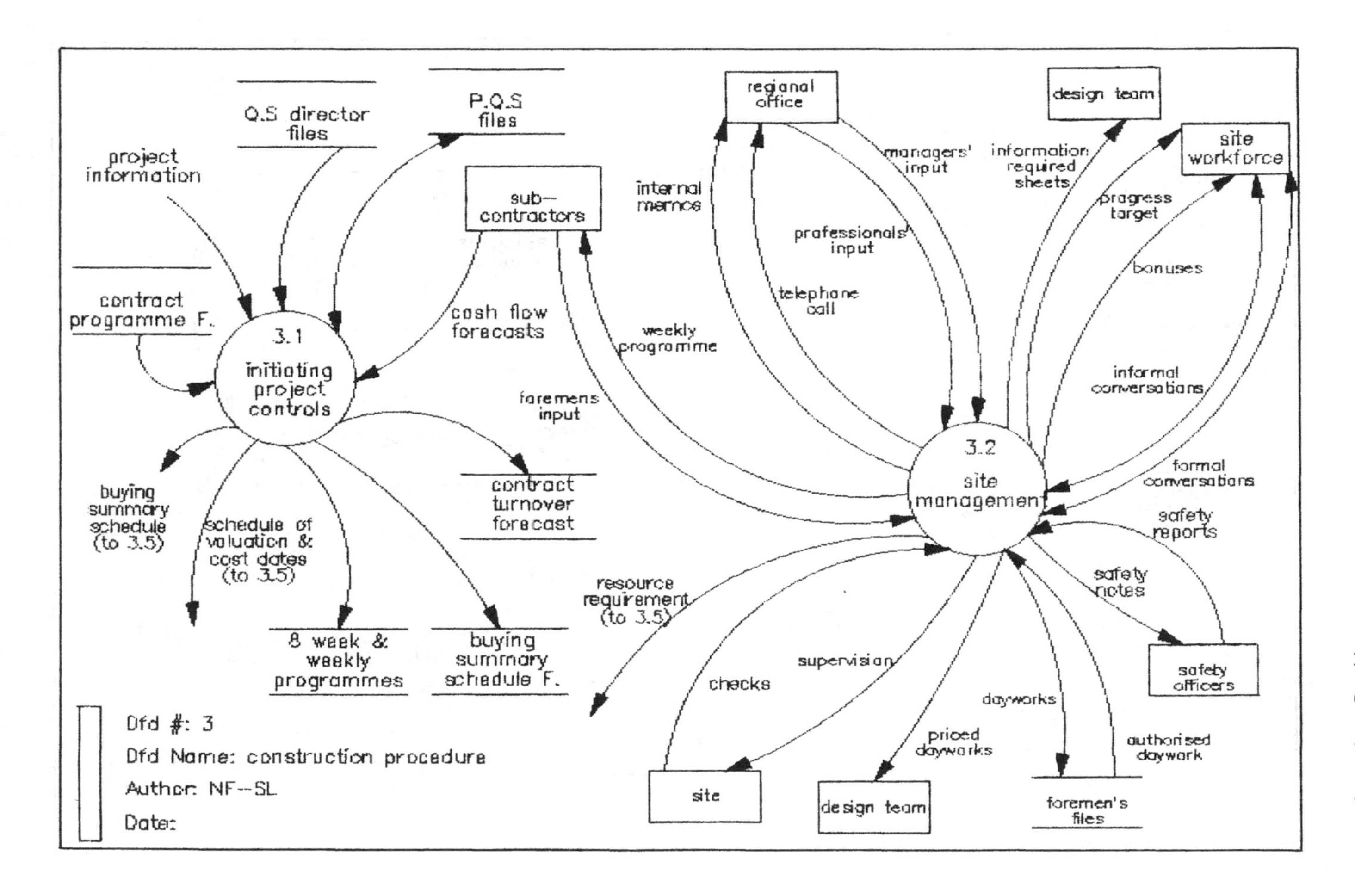
project information
Q.S director files
P.Q.S files
sub–contractors
regional office
design team
site workforce
contract programme F.
3.1 initiating project controls
3.2 site management
cash flow forecasts
internal memos
managers' input
information required sheets
professionals' input
progress target
telephone call
bonuses
weekly programme
informal conversations
foremens input
formal conversations
safety reports
buying summary schedule (to 3.5)
schedule of valuation & cost dates (to 3.5)
contract turnover forecast
safety notes
resource requirement (to 3.5)
8 week & weekly programmes
buying summary schedule F.
supervision
checks
safety officers
dayworks
priced dayworks
authorised daywork
site
design team
foremen's files
Dfd #: 3
Dfd Name: construction procedure
Author: NF--SL
Date:

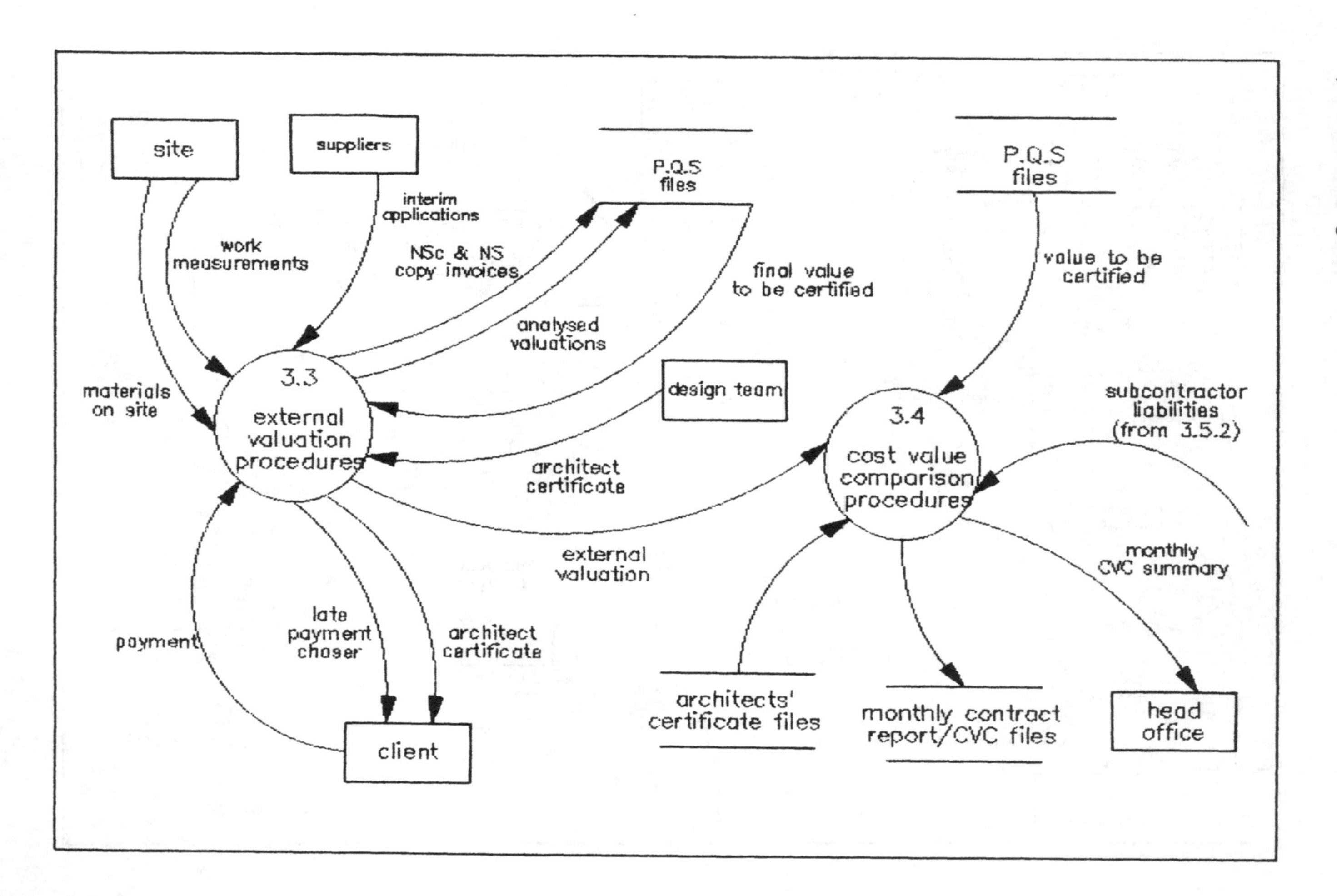
site
suppliers
P.Q.S files
P.Q.S files
interim applications
work measurements
NSc & NS copy invoices
final value to be certified
value to be certified
analysed valuations
design team
materials on site
3.3
external valuation procedures
3.4
cost value comparison procedures
subcontractor liabilities (from 3.5.2)
architect certificate
external valuation
monthly CVC summary
late payment chaser
architect certificate
payment
client
architects' certificate files
monthly contract report/CVC files
head office

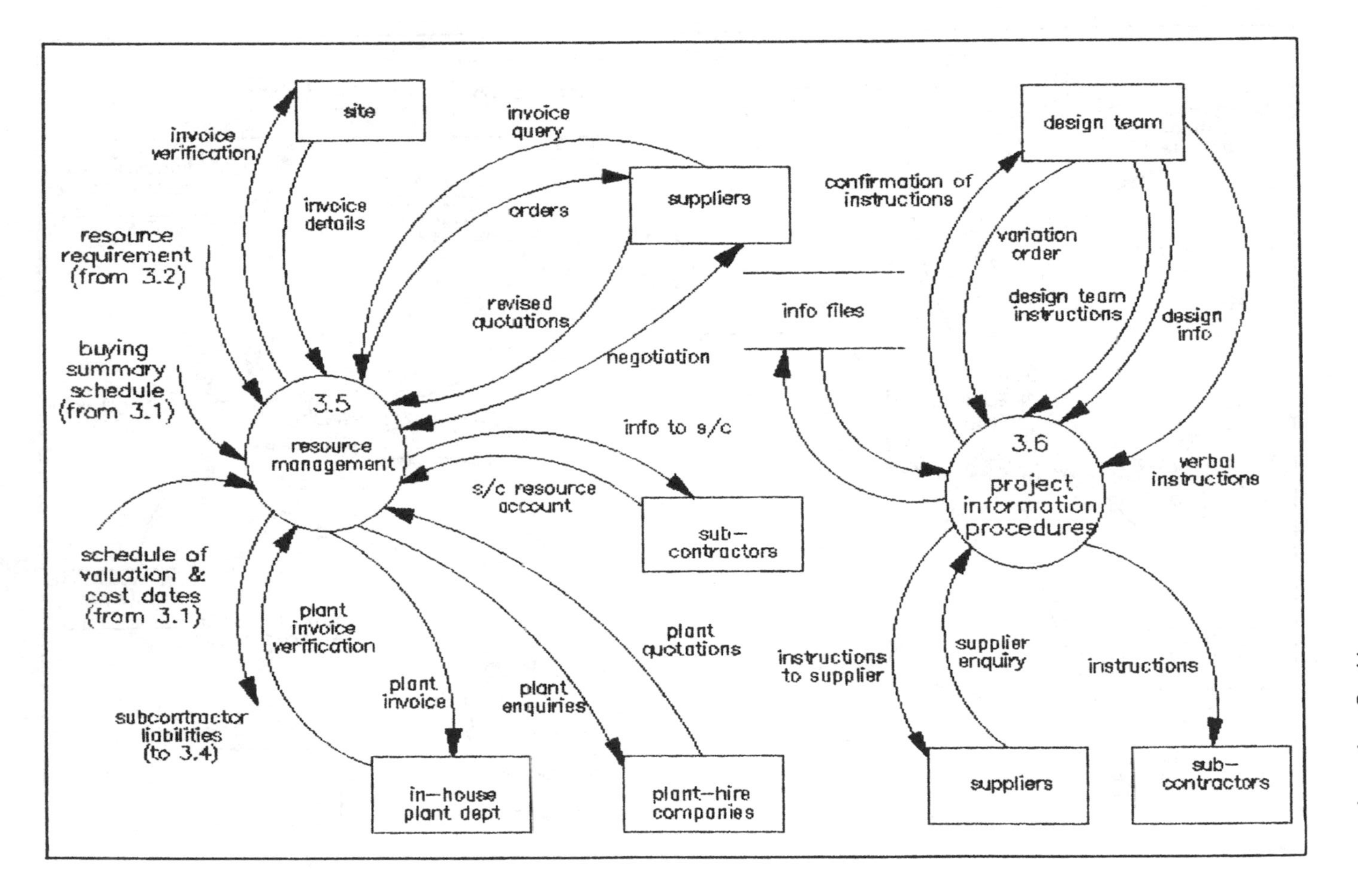

site
invoice query
design team
invoice verification
suppliers
confirmation of instructions
invoice details
orders
resource requirement (from 3.2)
variation order
revised quotations
info files
design team instructions
design info
buying summary schedule (from 3.1)
negotiation
3.5 resource management
info to s/c
3.6 project information procedures
verbal instructions
s/c resource account
sub-contractors
schedule of valuation & cost dates (from 3.1)
plant invoice verification
plant quotations
instructions to supplier
supplier enquiry
instructions
plant invoice
plant enquiries
subcontractor liabilities (to 3.4)
in-house plant dept
plant-hire companies
suppliers
sub-contractors

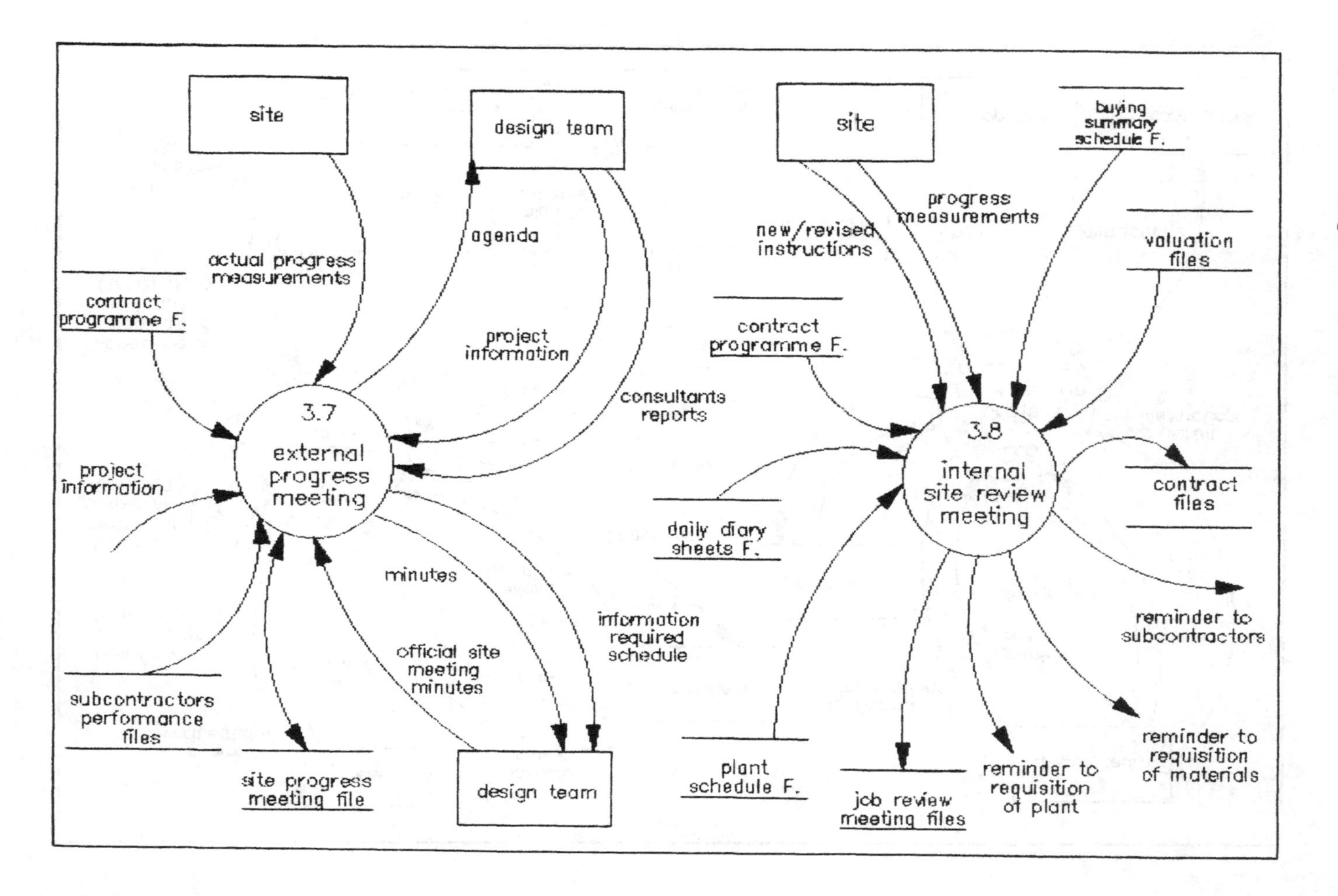
site
design team
site
buying summary schedule F.
actual progress measurements
agenda
new/revised instructions
progress measurements
valuation files
contract programme F.
project information
contract programme F.
consultants reports
3.7
external progress meeting
3.8
internal site review meeting
project information
contract files
daily diary sheets F.
minutes
information required schedule
reminder to subcontractors
official site meeting minutes
subcontractors performance files
reminder to requisition of materials
plant schedule F.
reminder to requisition of plant
site progress meeting file
design team
job review meeting files

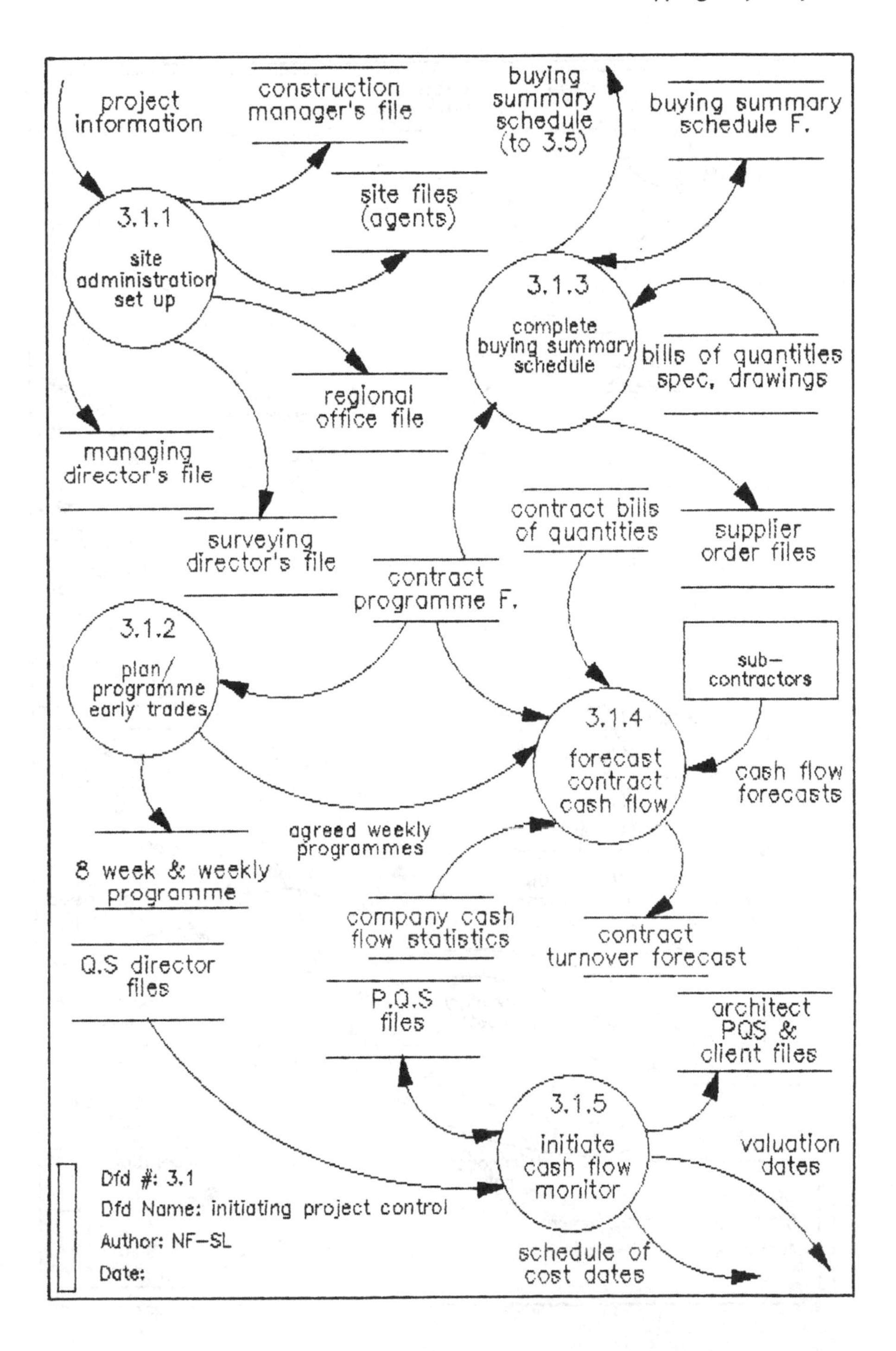

project information
construction manager's file
buying summary schedule (to 3.5)
buying summary schedule F.
site files (agents)
3.1.1
site administration set up
3.1.3
complete buying summary schedule
bills of quantities spec, drawings
regional office file
managing director's file
surveying director's file
contract bills of quantities
supplier order files
contract programme F.
3.1.2
plan/ programme early trades
sub-contractors
3.1.4
forecast contract cash flow
cash flow forecasts
agreed weekly programmes
8 week & weekly programme
company cash flow statistics
contract turnover forecast
Q.S director files
P.Q.S files
architect PQS & client files
3.1.5
initiate cash flow monitor
valuation dates
schedule of cost dates
Dfd #: 3.1
Dfd Name: initiating project control
Author: NF-SL
Date:

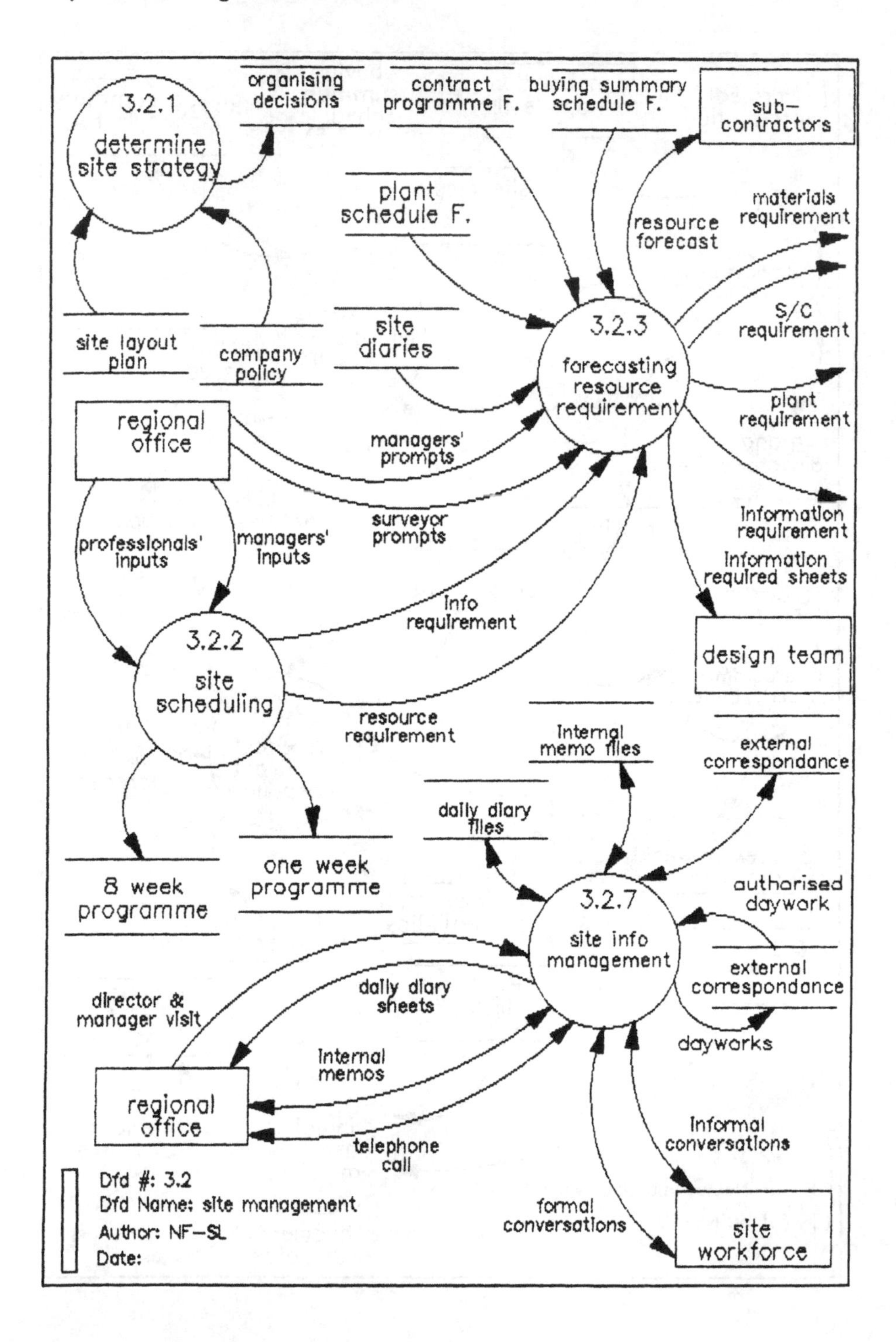
3.2.1
determine site strategy
organising decisions
contract programme F.
buying summary schedule F.
sub-contractors
plant schedule F.
resource forecast
materials requirement
S/C requirement
site layout plan
company policy
site diaries
3.2.3
forecasting resource requirement
plant requirement
regional office
managers' prompts
surveyor prompts
information requirement
information required sheets
professionals' inputs
managers' inputs
info requirement
3.2.2
site scheduling
design team
resource requirement
internal memo files
external correspondance
daily diary files
8 week programme
one week programme
3.2.7
site info management
authorised daywork
external correspondance
daily diary sheets
director & manager visit
dayworks
internal memos
regional office
telephone call
informal conversations
Dfd #: 3.2
Dfd Name: site management
Author: NF-SL
Date:
formal conversations
site workforce

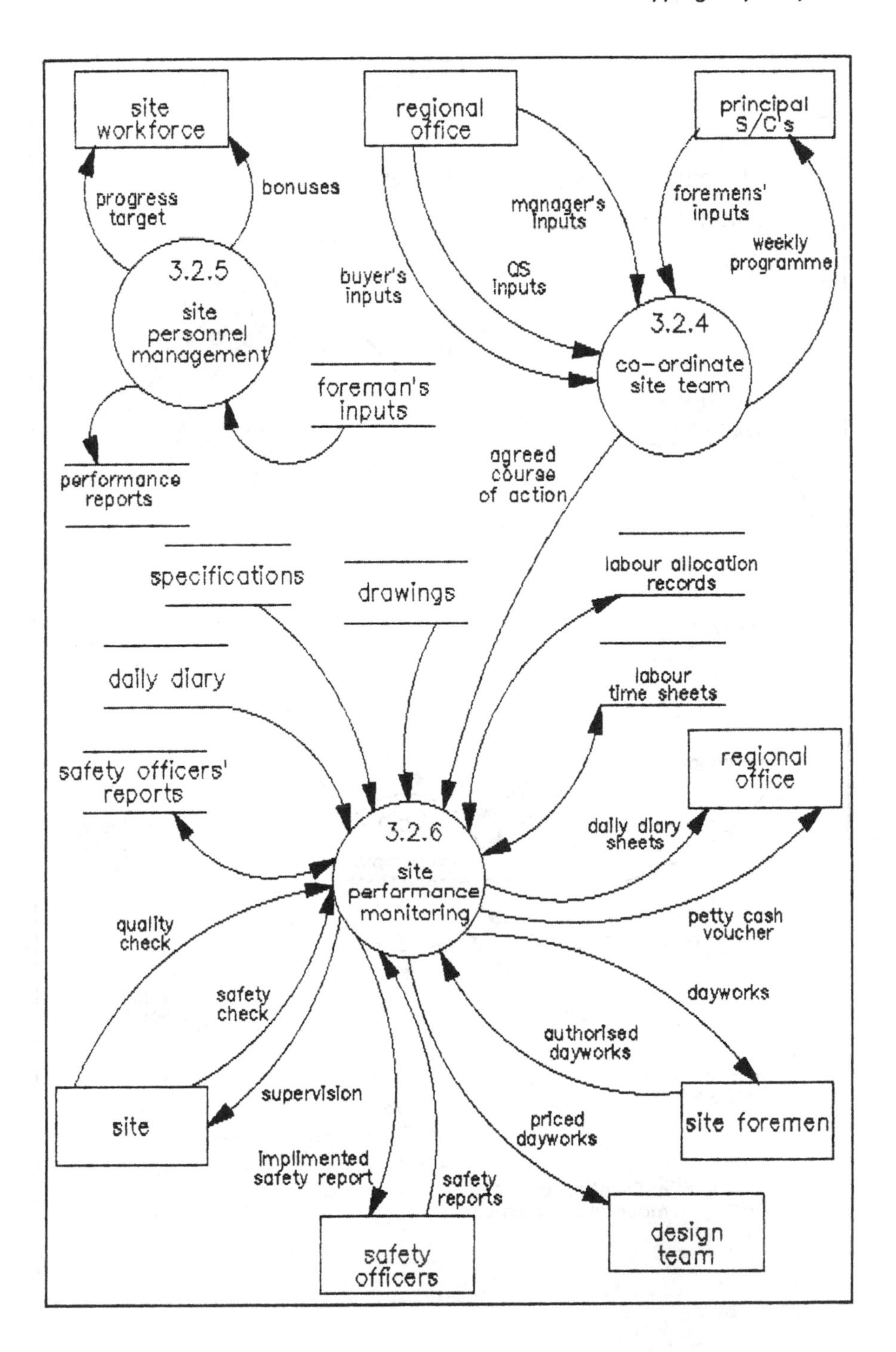
site workforce
regional office
principal S/C's
progress target
bonuses
manager's inputs
foremens' inputs
weekly programme
3.2.5
site personnel management
buyer's inputs
QS inputs
3.2.4
co-ordinate site team
foreman's inputs
performance reports
agreed course of action
specifications
drawings
labour allocation records
daily diary
labour time sheets
regional office
safety officers' reports
3.2.6
site performance monitoring
daily diary sheets
petty cash voucher
quality check
safety check
dayworks
authorised dayworks
supervision
site
site foremen
priced dayworks
implimented safety report
safety reports
design team
safety officers

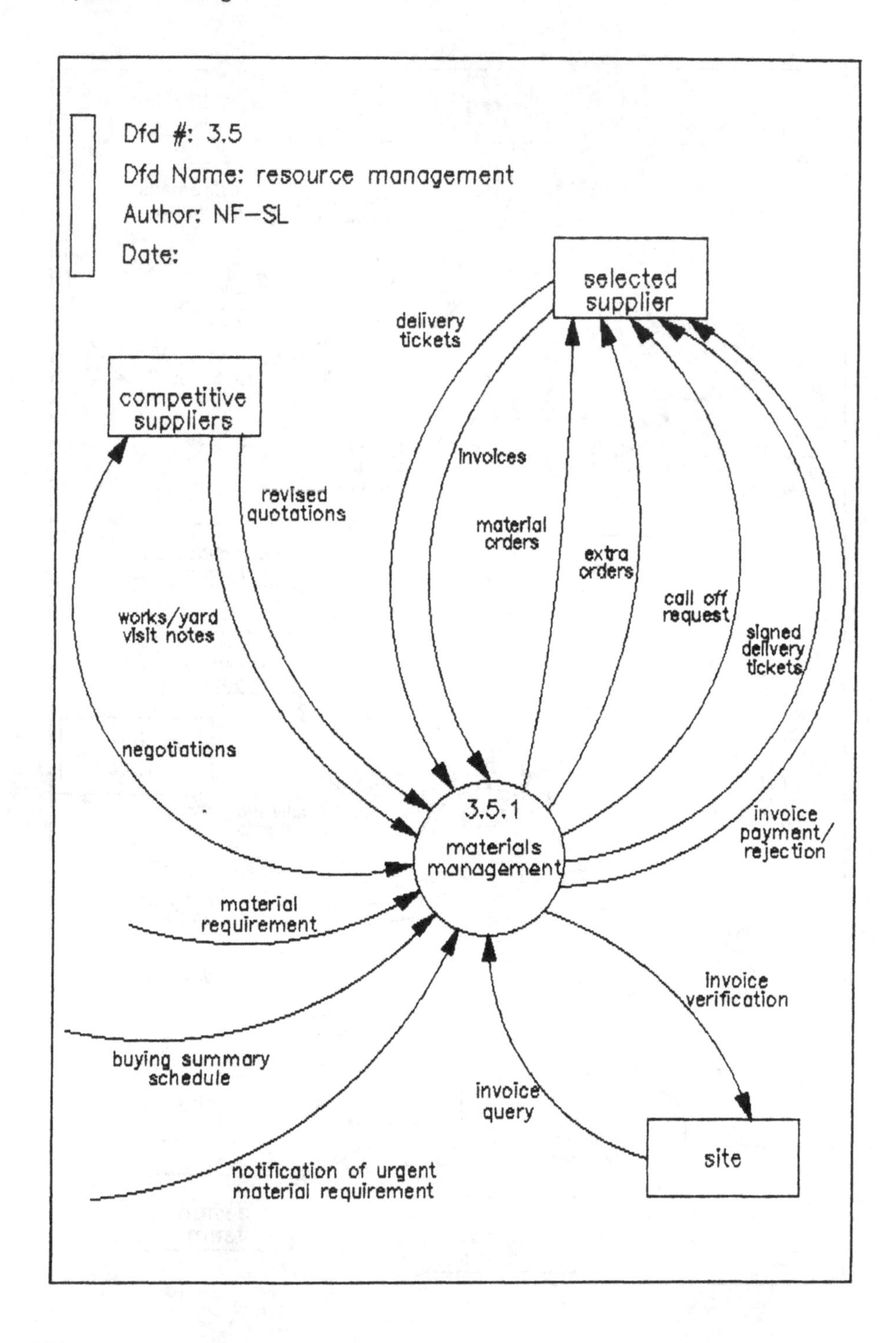
Dfd #: 3.5
Dfd Name: resource management
Author: NF-SL
Date:
selected supplier
competitive suppliers
delivery tickets
invoices
material orders
extra orders
call off request
signed delivery tickets
revised quotations
works/yard visit notes
negotiations
3.5.1
materials management
invoice payment/ rejection
material requirement
buying summary schedule
notification of urgent material requirement
invoice verification
invoice query
site

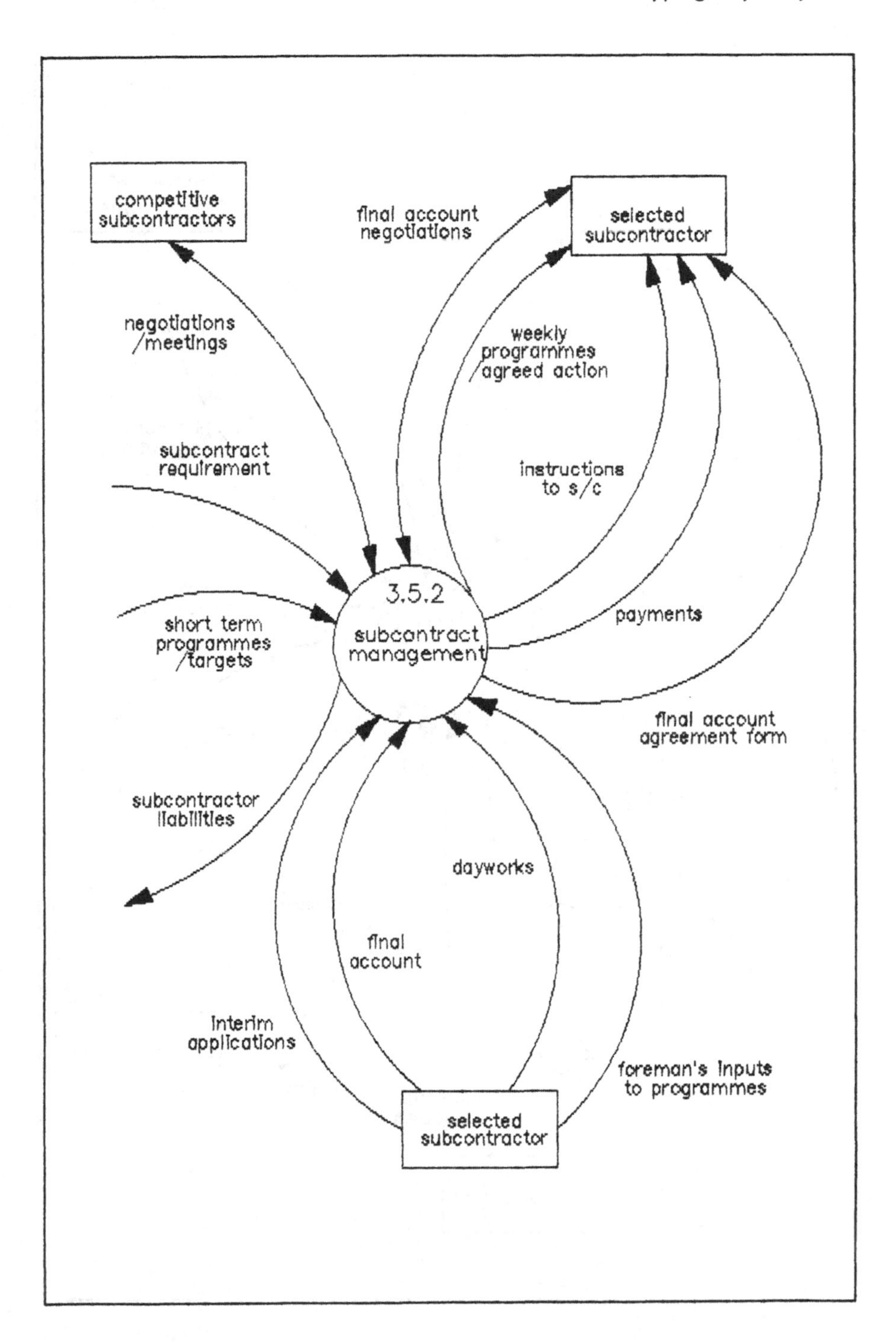
competitive subcontractors
selected subcontractor
final account negotiations
negotiations /meetings
weekly programmes /agreed action
subcontract requirement
instructions to s/c
3.5.2
subcontract management
short term programmes /targets
payments
final account agreement form
subcontractor liabilities
dayworks
final account
interim applications
foreman's inputs to programmes
selected subcontractor

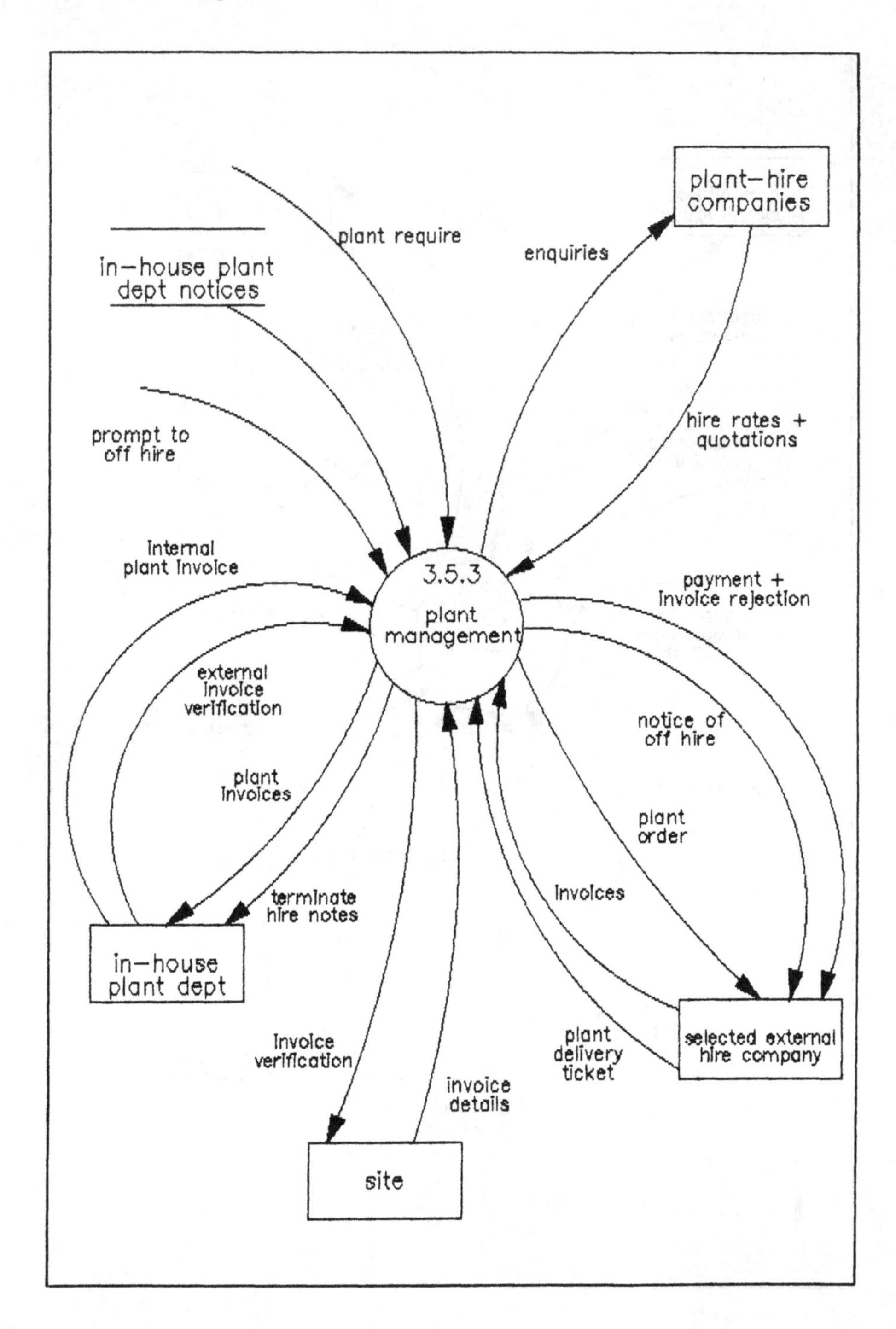
plant-hire companies
plant require
enquiries
in-house plant dept notices
prompt to off hire
hire rates + quotations
internal plant invoice
3.5.3 plant management
payment + invoice rejection
external invoice verification
notice of off hire
plant invoices
plant order
invoices
terminate hire notes
in-house plant dept
plant delivery ticket
selected external hire company
invoice verification
invoice details
site

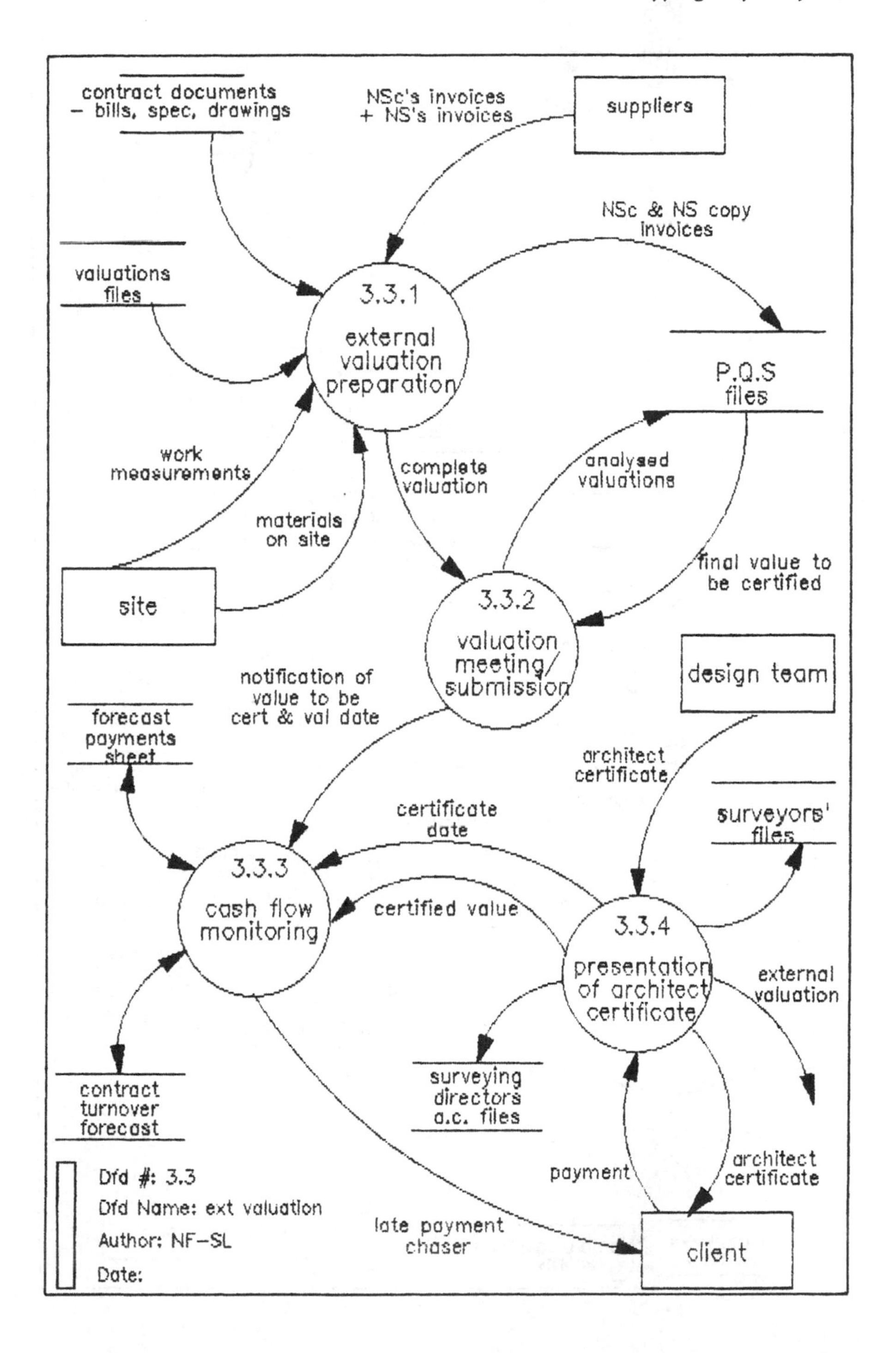
contract documents
– bills, spec, drawings
NSc's invoices
+ NS's invoices
suppliers
NSc & NS copy
invoices
valuations
files
3.3.1
external
valuation
preparation
P.Q.S
files
work
measurements
complete
valuation
analysed
valuations
materials
on site
final value to
be certified
site
3.3.2
valuation
meeting/
submission
design team
notification of
value to be
cert & val date
forecast
payments
sheet
architect
certificate
surveyors'
files
certificate
date
3.3.3
cash flow
monitoring
certified value
3.3.4
presentation
of architect
certificate
external
valuation
surveying
directors
a.c. files
contract
turnover
forecast
Dfd #: 3.3
Dfd Name: ext valuation
Author: NF–SL
Date:
payment
architect
certificate
late payment
chaser
client

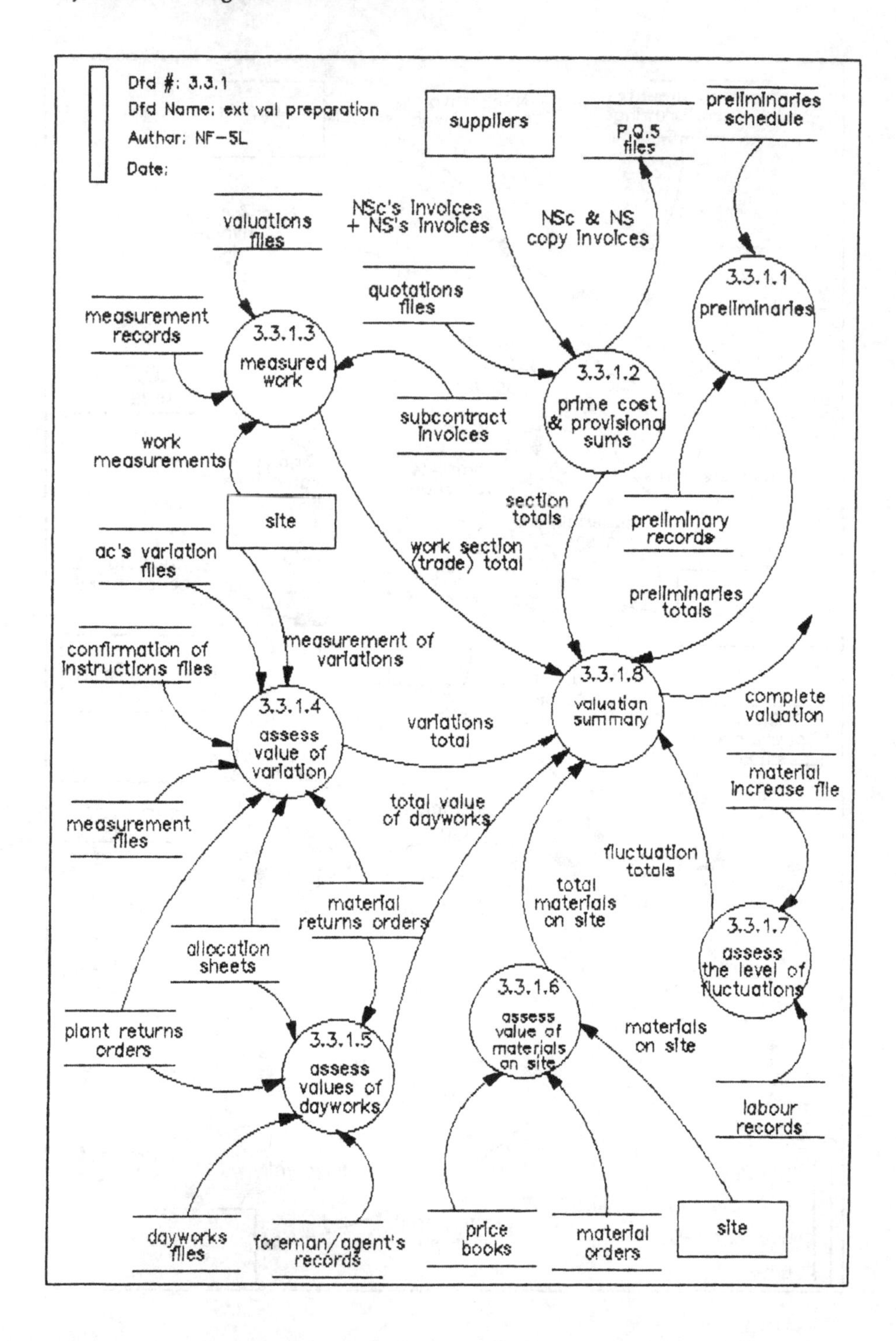

Dfd #: 3.3.1
Dfd Name: ext val preparation
Author: NF-5L
Date:
suppliers
P.O.S files
preliminaries schedule
valuations files
NSc's invoices + NS's invoices
NSc & NS copy invoices
3.3.1.1 preliminaries
quotations files
measurement records
3.3.1.3 measured work
3.3.1.2 prime cost & provisional sums
subcontract invoices
work measurements
site
section totals
preliminary records
ac's variation files
work section (trade) total
preliminaries totals
confirmation of instructions files
measurement of variations
3.3.1.8 valuation summary
complete valuation
3.3.1.4 assess value of variation
variations total
material increase file
measurement files
total value of dayworks
fluctuation totals
total materials on site
material returns orders
3.3.1.7 assess the level of fluctuations
allocation sheets
3.3.1.6 assess value of materials on site
materials on site
plant returns orders
3.3.1.5 assess values of dayworks
labour records
dayworks files
foreman/agent's records
price books
material orders
site

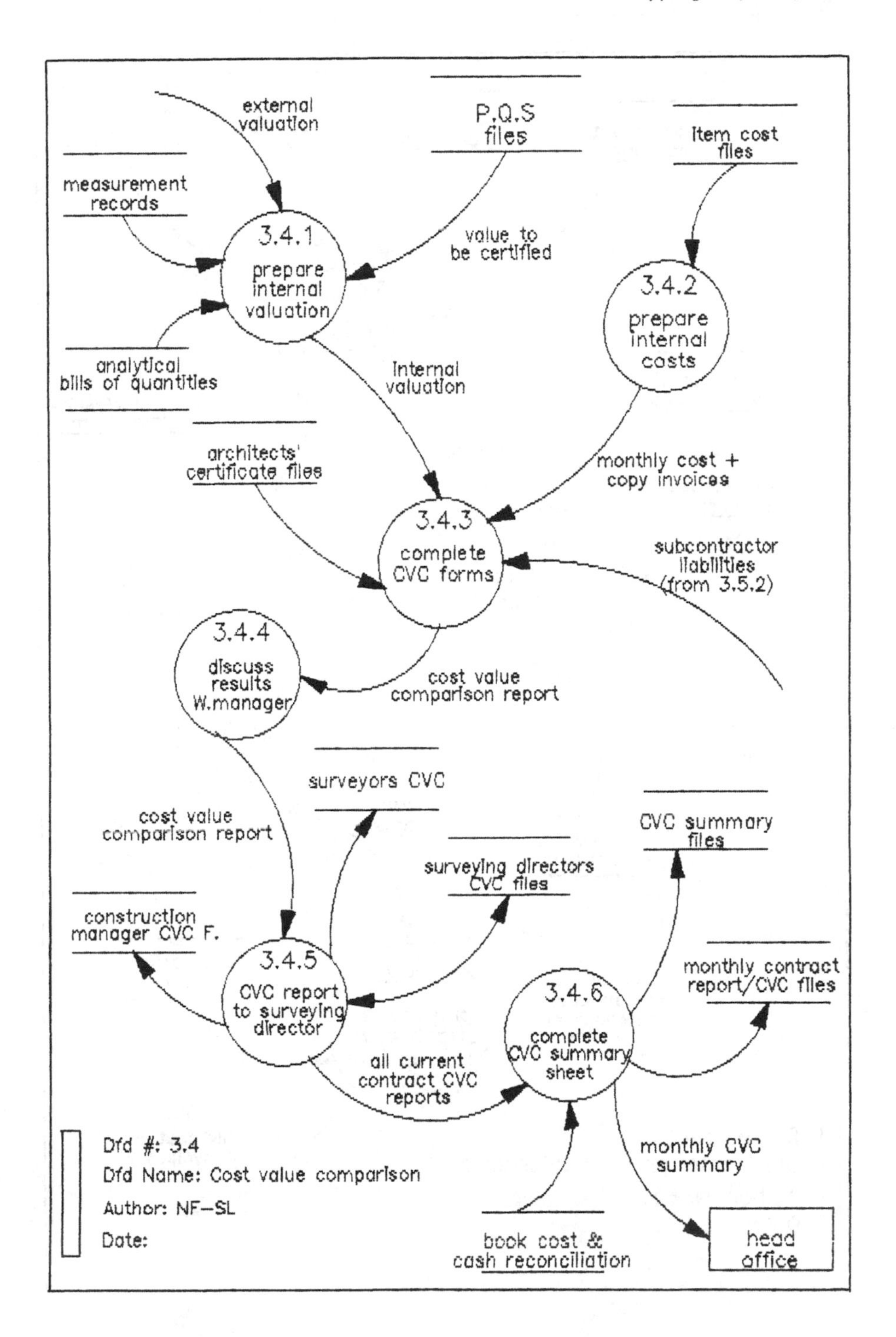
external valuation
P.Q.S files
Item cost files
measurement records
3.4.1 prepare internal valuation
value to be certified
3.4.2 prepare internal costs
analytical bills of quantities
internal valuation
architects' certificate files
monthly cost + copy invoices
3.4.3 complete CVC forms
subcontractor liabilities (from 3.5.2)
3.4.4 discuss results W.manager
cost value comparison report
surveyors CVC
cost value comparison report
CVC summary files
surveying directors CVC files
construction manager CVC F.
3.4.5 CVC report to surveying director
monthly contract report/CVC files
3.4.6 complete CVC summary sheet
all current contract CVC reports
Dfd #: 3.4
Dfd Name: Cost value comparison
Author: NF-SL
Date:
monthly CVC summary
book cost & cash reconciliation
head office

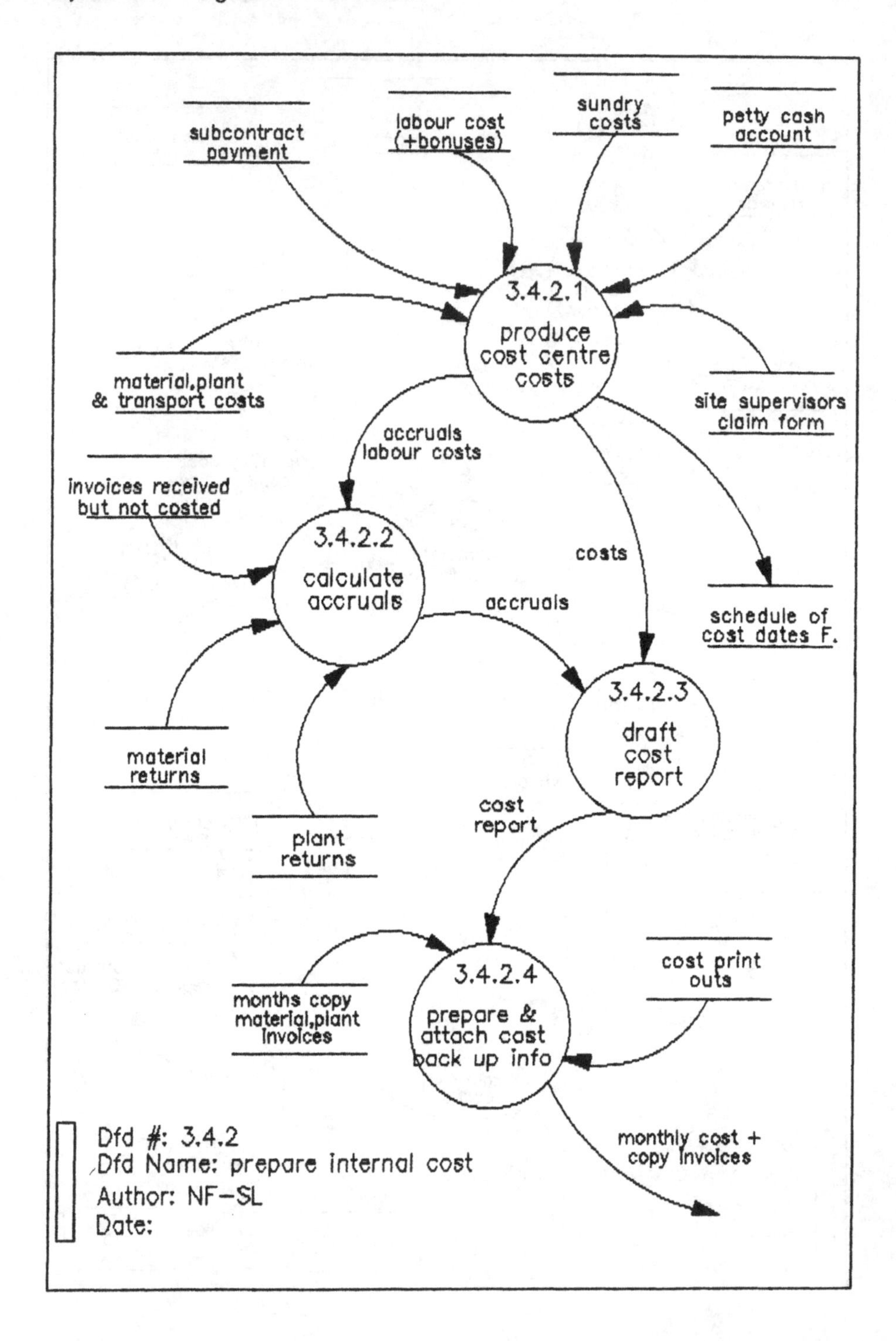
subcontract payment
labour cost (+bonuses)
sundry costs
petty cash account
3.4.2.1
produce cost centre costs
material,plant & transport costs
site supervisors claim form
accruals labour costs
invoices received but not costed
costs
3.4.2.2
calculate accruals
accruals
schedule of cost dates F.
3.4.2.3
draft cost report
material returns
plant returns
cost report
3.4.2.4
prepare & attach cost back up info
months copy material,plant invoices
cost print outs
monthly cost + copy invoices
Dfd #: 3.4.2
Dfd Name: prepare internal cost
Author: NF-SL
Date:

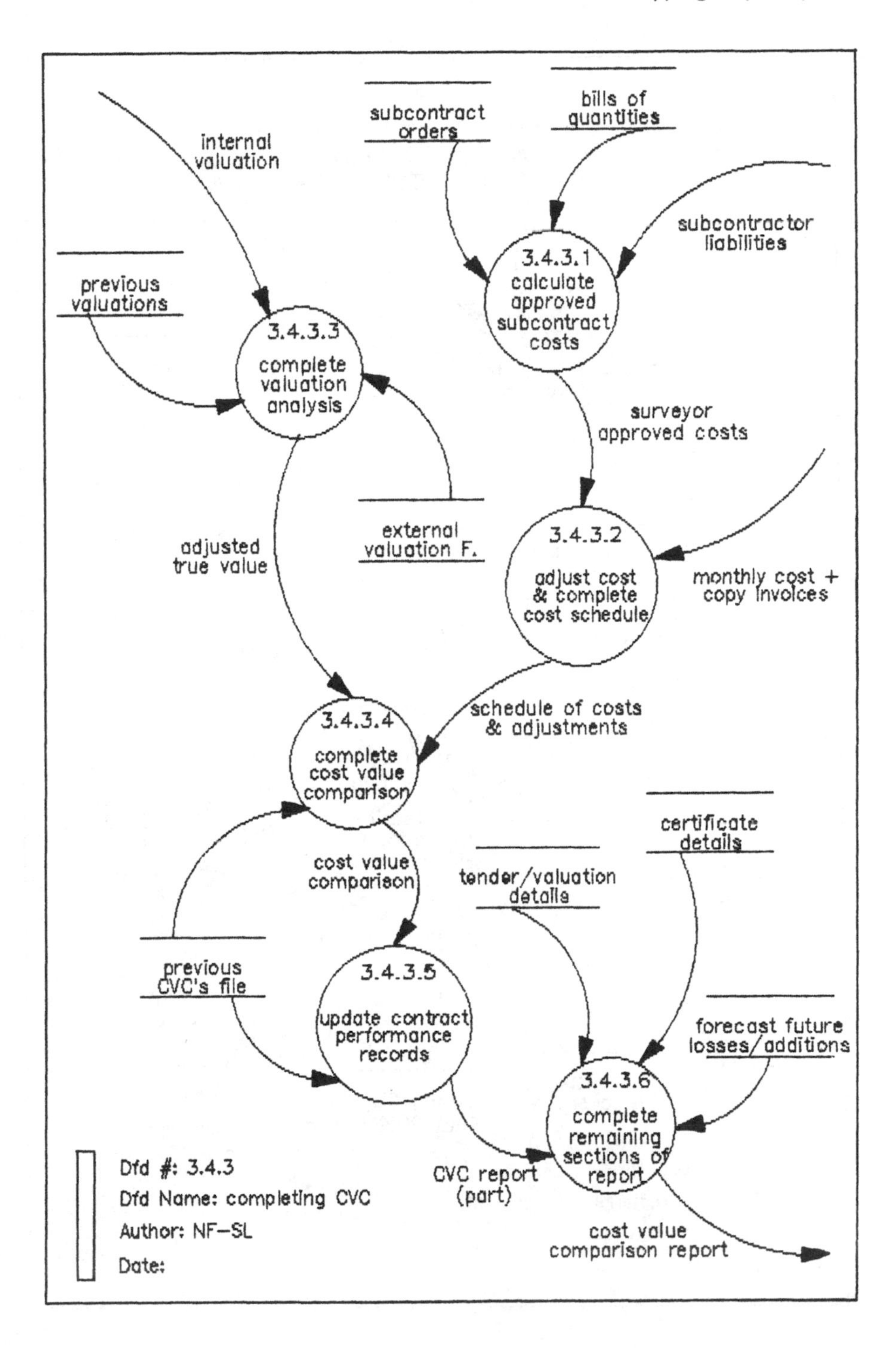
internal valuation
subcontract orders
bills of quantities
subcontractor liabilities
previous valuations
3.4.3.1 calculate approved subcontract costs
3.4.3.3 complete valuation analysis
surveyor approved costs
external valuation F.
adjusted true value
3.4.3.2 adjust cost & complete cost schedule
monthly cost + copy invoices
schedule of costs & adjustments
3.4.3.4 complete cost value comparison
certificate details
cost value comparison
tender/valuation details
previous CVC's file
3.4.3.5 update contract performance records
forecast future losses/additions
3.4.3.6 complete remaining sections of report
Dfd #: 3.4.3
Dfd Name: completing CVC
Author: NF–SL
Date:
CVC report (part)
cost value comparison report

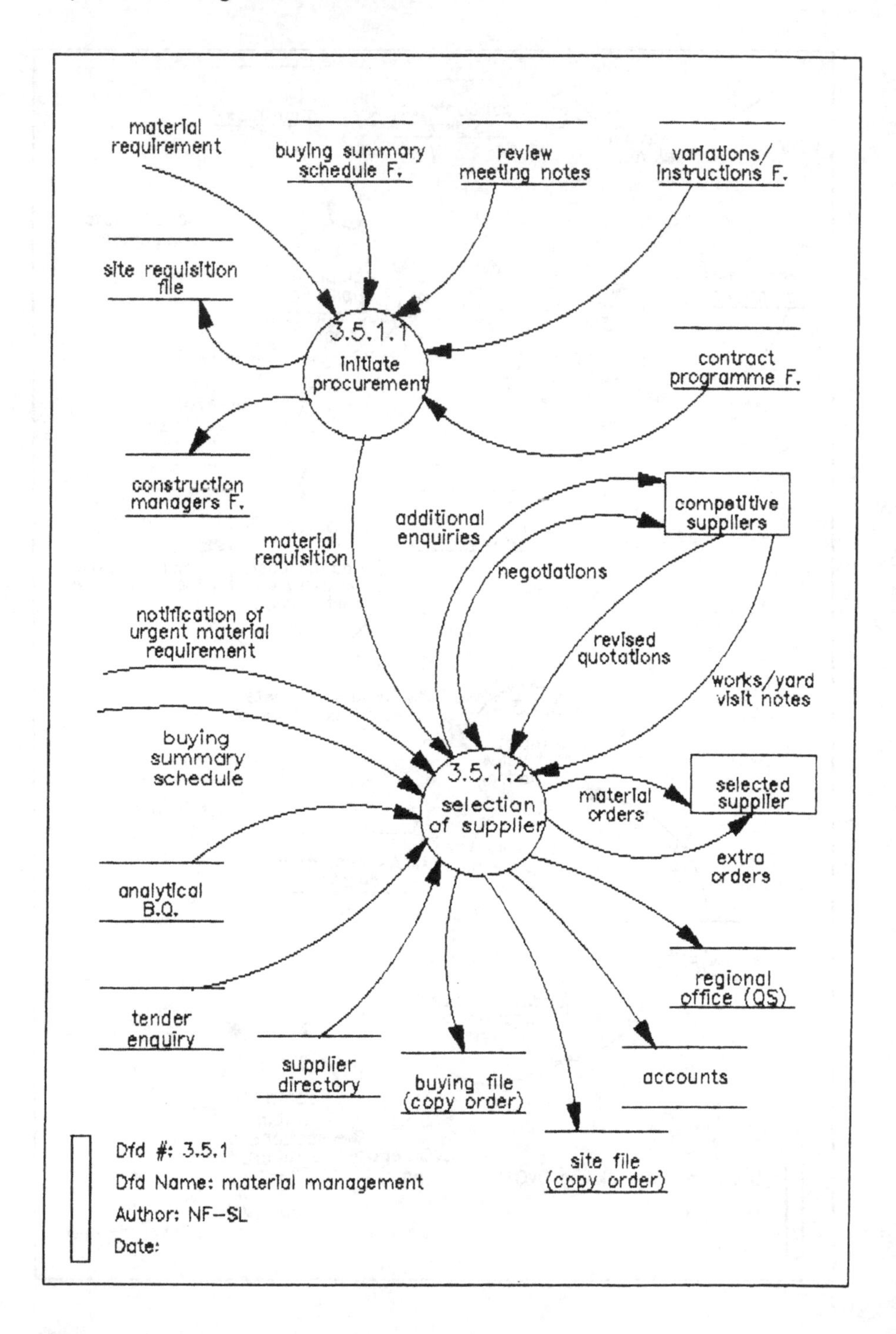
material requirement
buying summary schedule F.
review meeting notes
variations/ instructions F.
site requisition file
3.5.1.1
initiate procurement
contract programme F.
construction managers F.
material requisition
additional enquiries
competitive suppliers
negotiations
notification of urgent material requirement
revised quotations
works/yard visit notes
buying summary schedule
3.5.1.2
selection of supplier
material orders
selected supplier
extra orders
analytical B.Q.
regional office (QS)
tender enquiry
supplier directory
buying file (copy order)
accounts
site file (copy order)
Dfd #: 3.5.1
Dfd Name: material management
Author: NF-SL
Date:

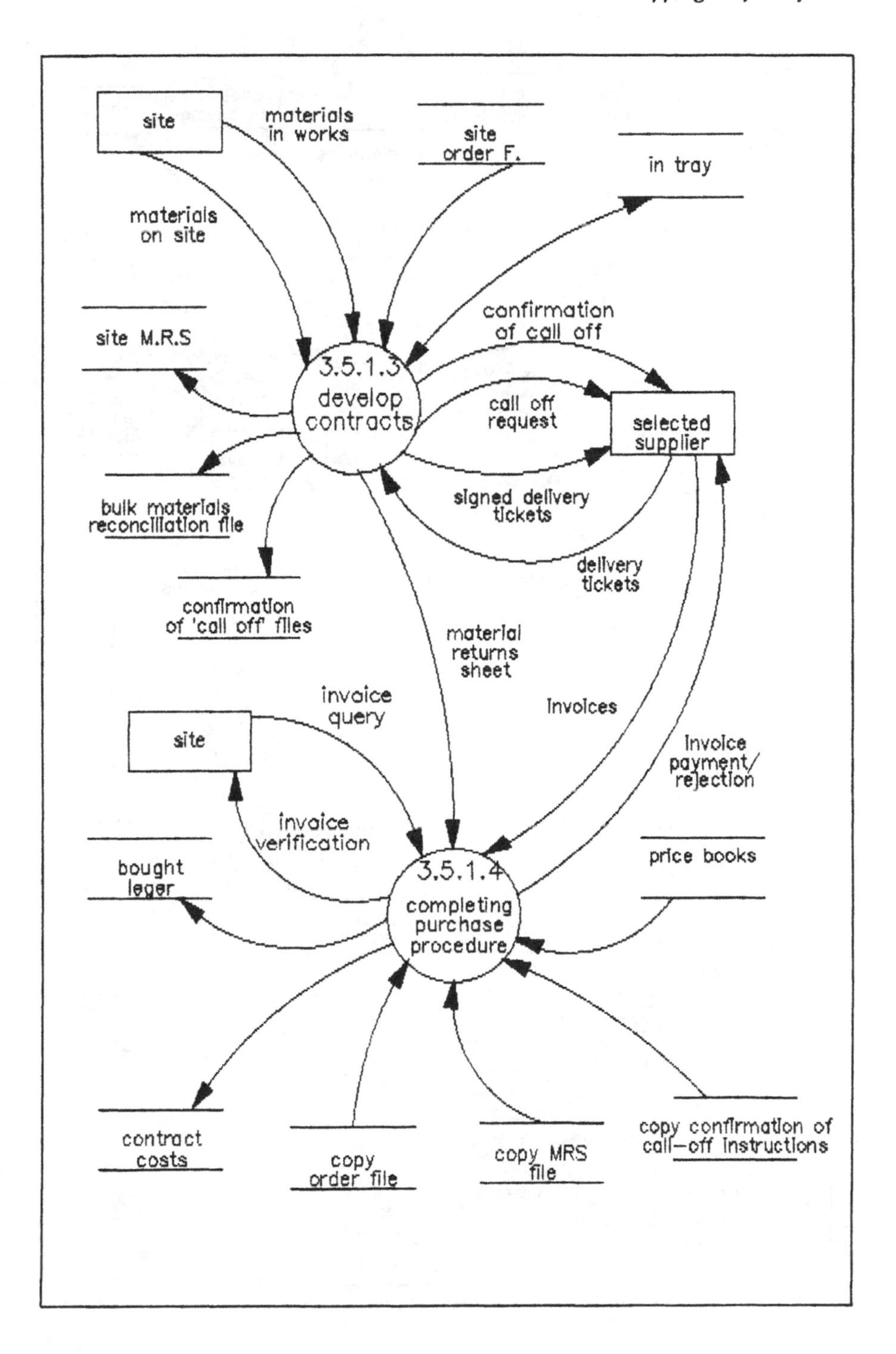
site
materials in works
site order F.
in tray
materials on site
site M.R.S
confirmation of call off
3.5.1.3 develop contracts
call off request
selected supplier
bulk materials reconciliation file
signed delivery tickets
delivery tickets
confirmation of 'call off' files
material returns sheet
invoice query
invoices
site
invoice payment/ rejection
invoice verification
bought leger
3.5.1.4 completing purchase procedure
price books
contract costs
copy order file
copy MRS file
copy confirmation of call-off instructions

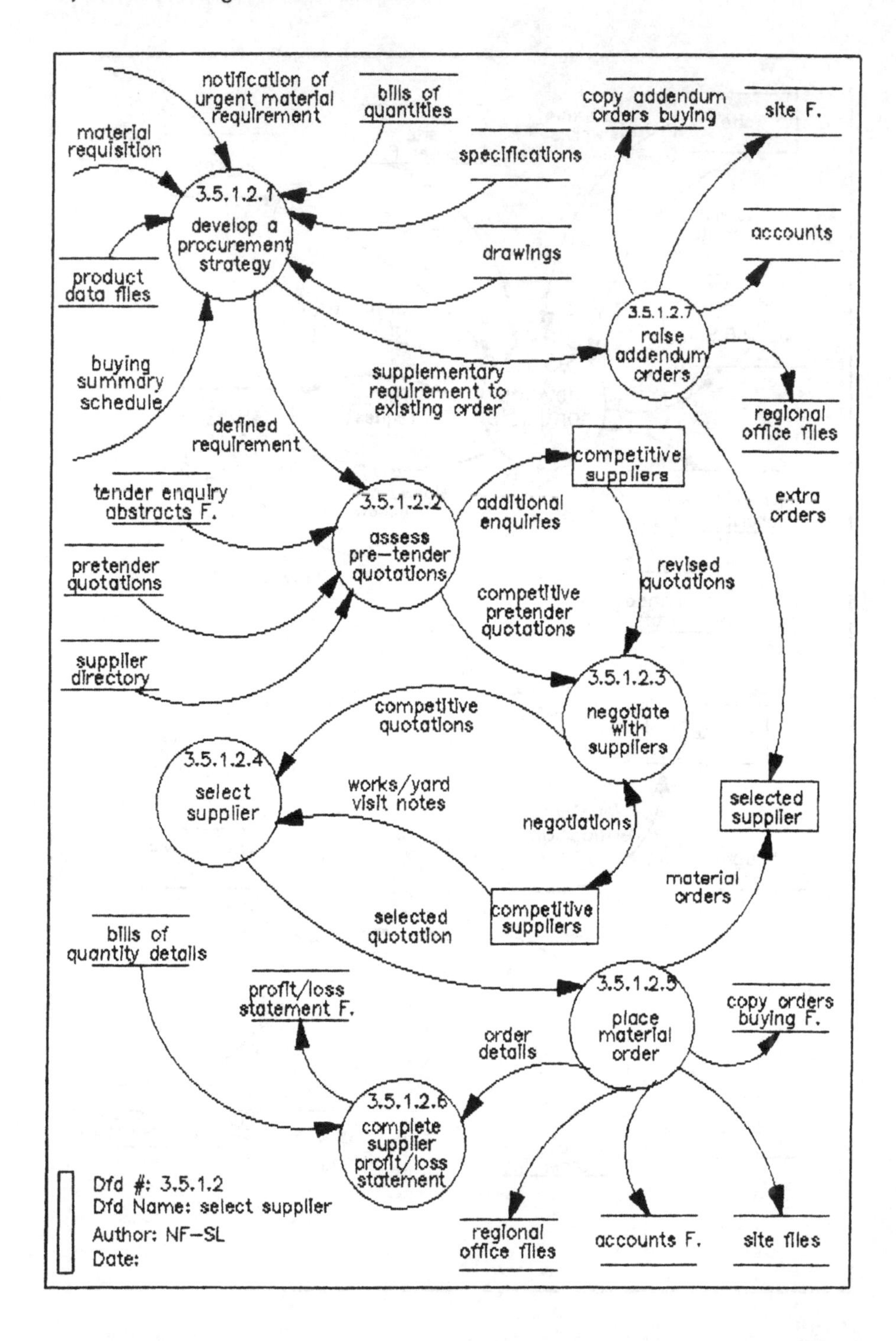
notification of urgent material requirement
bills of quantities
material requisition
specifications
3.5.1.2.1 develop a procurement strategy
drawings
product data files
buying summary schedule
copy addendum orders buying
site F.
accounts
3.5.1.2.7 raise addendum orders
regional office files
supplementary requirement to existing order
defined requirement
competitive suppliers
tender enquiry abstracts F.
3.5.1.2.2 assess pre-tender quotations
additional enquiries
extra orders
pretender quotations
revised quotations
competitive pretender quotations
supplier directory
3.5.1.2.3 negotiate with suppliers
competitive quotations
3.5.1.2.4 select supplier
works/yard visit notes
negotiations
selected supplier
material orders
competitive suppliers
selected quotation
bills of quantity details
profit/loss statement F.
3.5.1.2.5 place material order
copy orders buying F.
order details
3.5.1.2.6 complete supplier profit/loss statement
Dfd #: 3.5.1.2
Dfd Name: select supplier
Author: NF–SL
Date:
regional office files
accounts F.
site files

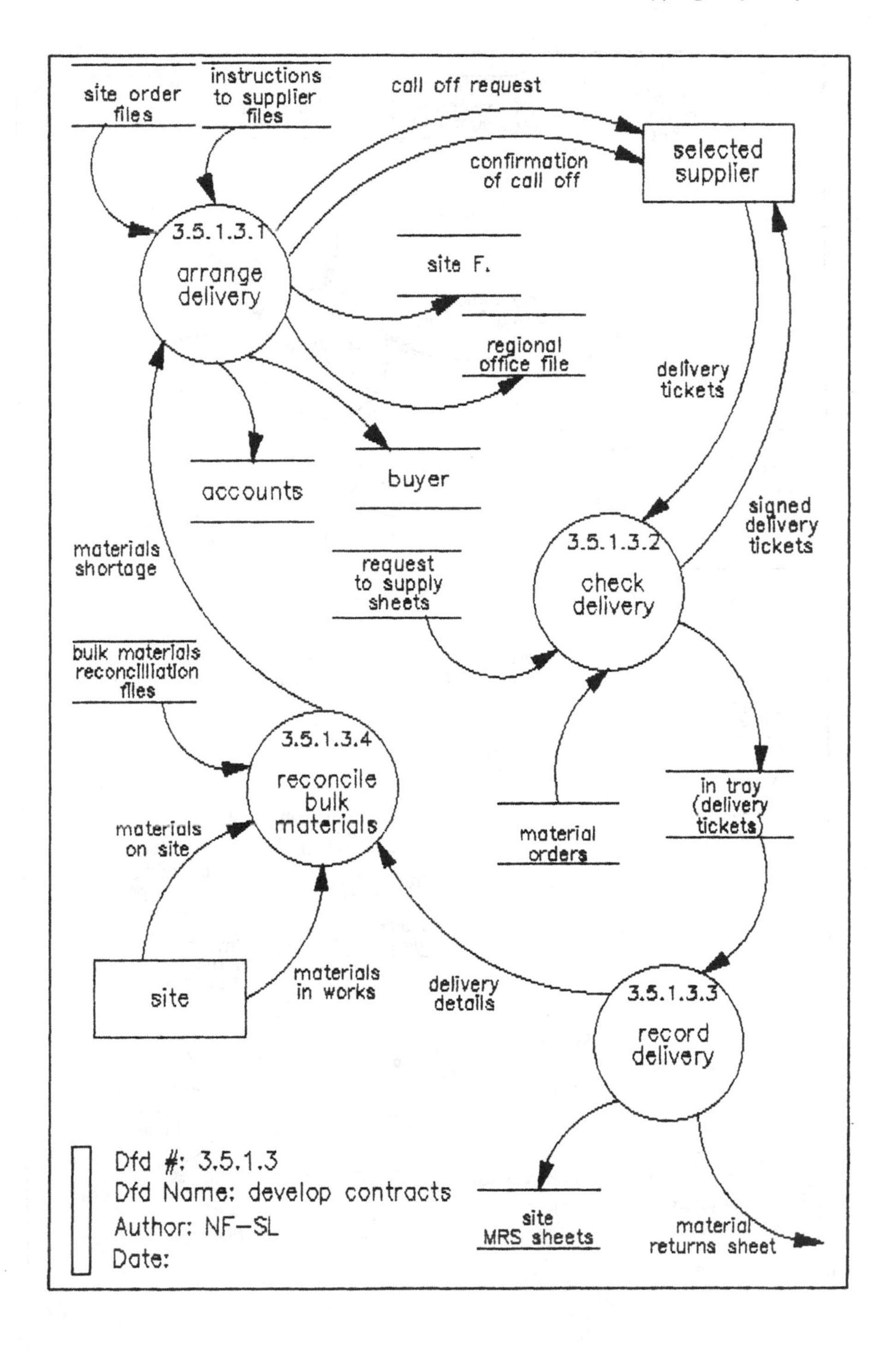

site order files
instructions to supplier files
call off request
confirmation of call off
selected supplier
3.5.1.3.1
arrange delivery
site F.
regional office file
delivery tickets
accounts
buyer
signed delivery tickets
materials shortage
request to supply sheets
3.5.1.3.2
check delivery
bulk materials reconciliation files
3.5.1.3.4
reconcile bulk materials
materials on site
in tray (delivery tickets)
material orders
site
materials in works
delivery details
3.5.1.3.3
record delivery
Dfd #: 3.5.1.3
Dfd Name: develop contracts
Author: NF–SL
Date:
site MRS sheets
material returns sheet

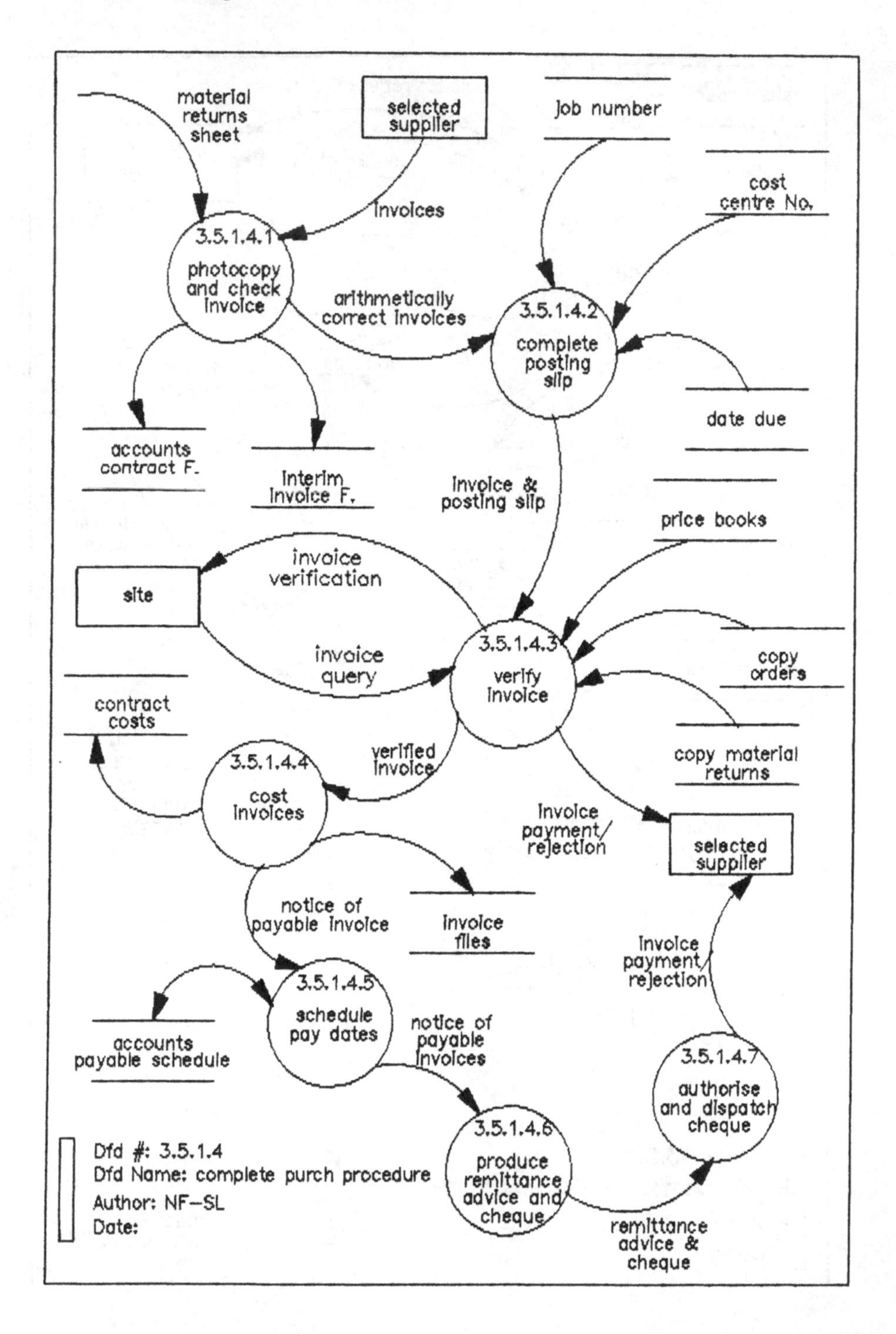

material returns sheet
selected supplier
invoices
job number
cost centre No.
3.5.1.4.1 photocopy and check invoice
arithmetically correct invoices
3.5.1.4.2 complete posting slip
date due
accounts contract F.
interim invoice F.
invoice & posting slip
price books
site
invoice verification
invoice query
3.5.1.4.3 verify invoice
copy orders
copy material returns
contract costs
verified invoice
3.5.1.4.4 cost invoices
invoice payment/ rejection
selected supplier
notice of payable invoice
invoice files
invoice payment/ rejection
3.5.1.4.5 schedule pay dates
accounts payable schedule
notice of payable invoices
3.5.1.4.7 authorise and dispatch cheque
3.5.1.4.6 produce remittance advice and cheque
remittance advice & cheque
Dfd #: 3.5.1.4
Dfd Name: complete purch procedure
Author: NF-SL
Date:

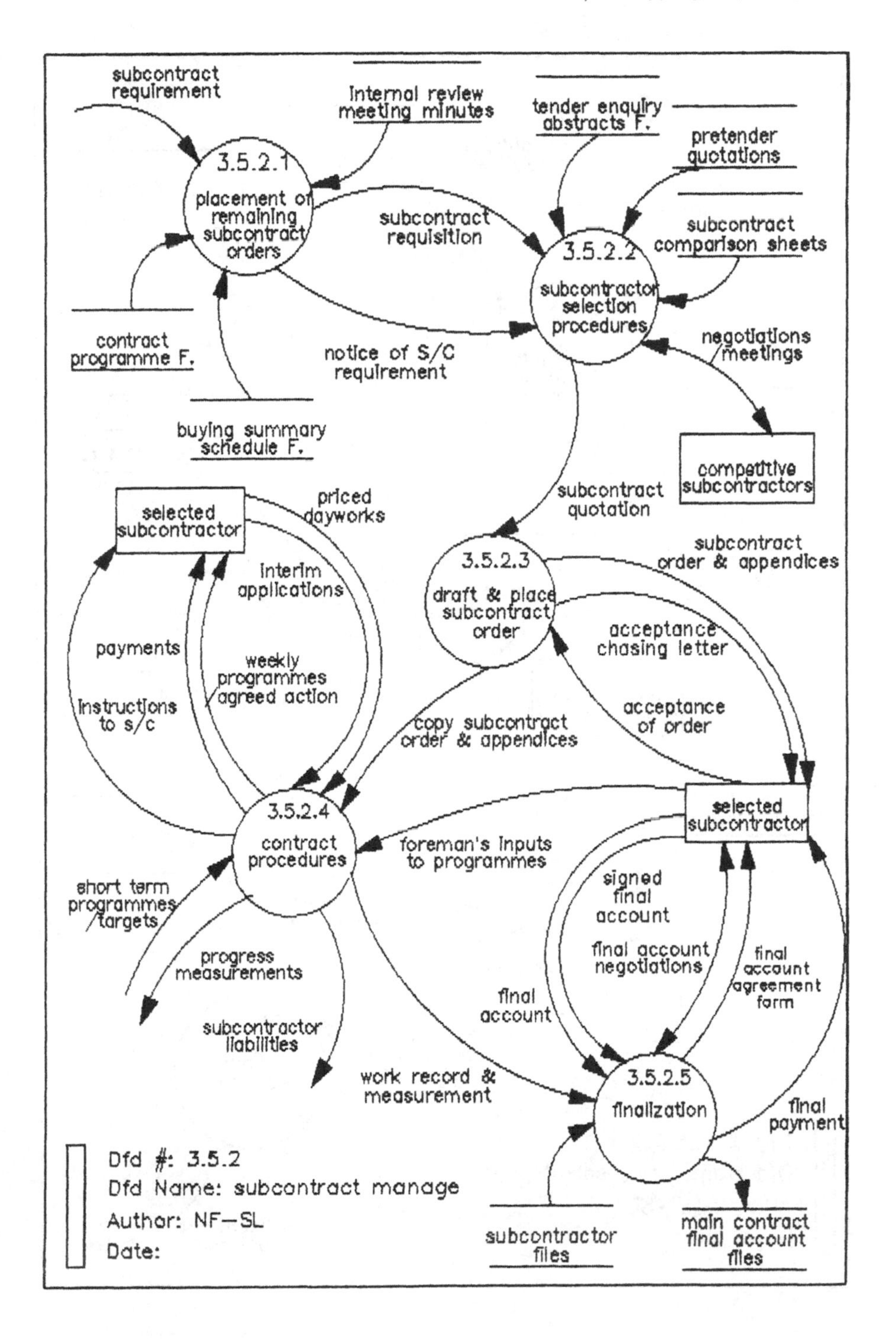
subcontract requirement
internal review meeting minutes
tender enquiry abstracts F.
pretender quotations
3.5.2.1
placement of remaining subcontract orders
subcontract requisition
subcontract comparison sheets
3.5.2.2
subcontractor selection procedures
contract programme F.
notice of S/C requirement
negotiations /meetings
buying summary schedule F.
competitive subcontractors
selected subcontractor
priced dayworks
subcontract quotation
subcontract order & appendices
3.5.2.3
draft & place subcontract order
interim applications
payments
weekly programmes /agreed action
acceptance chasing letter
instructions to s/c
acceptance of order
copy subcontract order & appendices
selected subcontractor
3.5.2.4
contract procedures
foreman's inputs to programmes
short term programmes /targets
signed final account
final account negotiations
final account agreement form
progress measurements
final account
subcontractor liabilities
work record & measurement
3.5.2.5
finalization
final payment
Dfd #: 3.5.2
Dfd Name: subcontract manage
Author: NF—SL
Date:
subcontractor files
main contract final account files

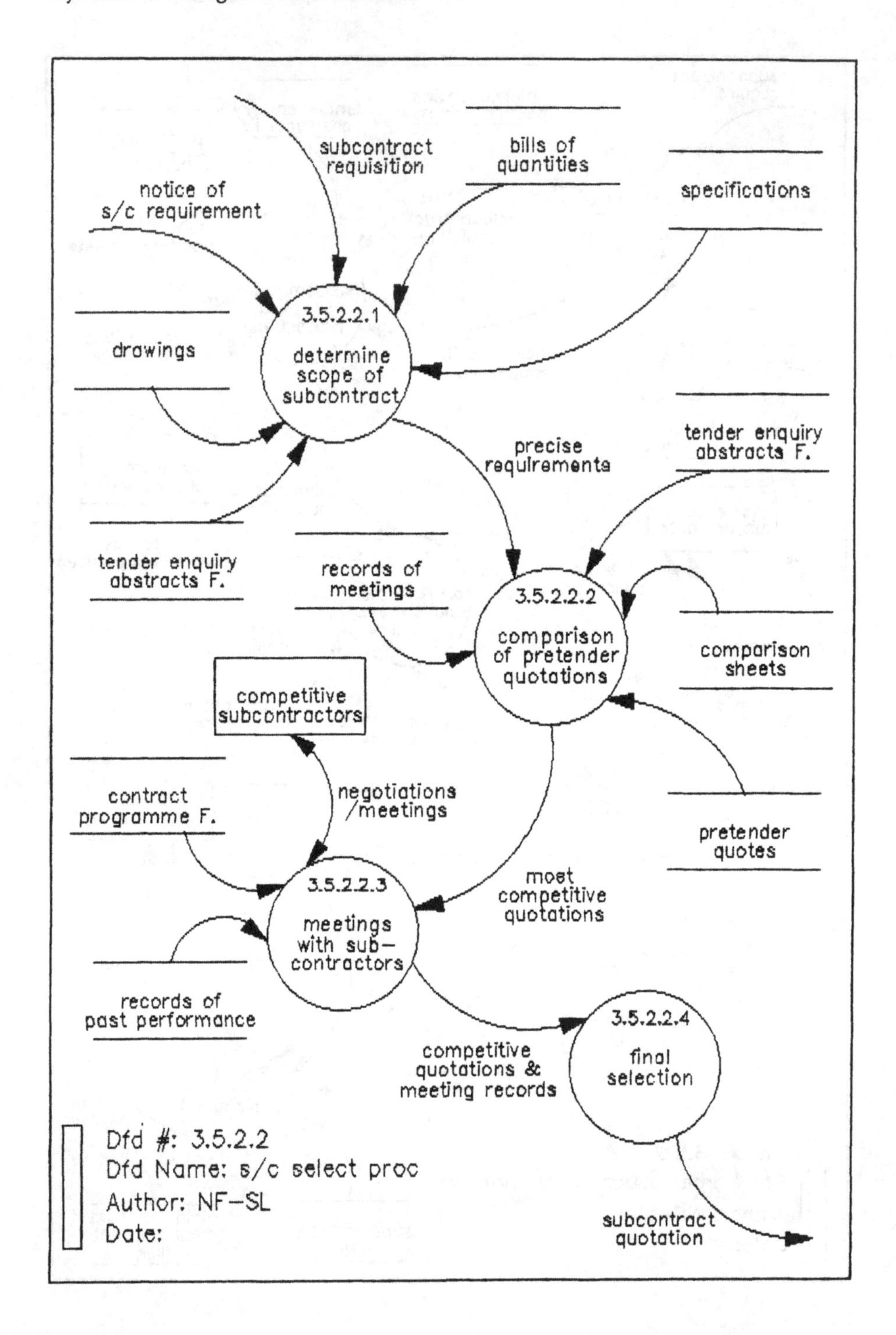
subcontract requisition
bills of quantities
specifications
notice of s/c requirement
drawings
3.5.2.2.1 determine scope of subcontract
precise requirements
tender enquiry abstracts F.
tender enquiry abstracts F.
records of meetings
3.5.2.2.2 comparison of pretender quotations
comparison sheets
competitive subcontractors
pretender quotes
contract programme F.
negotiations /meetings
most competitive quotations
3.5.2.2.3 meetings with sub-contractors
records of past performance
competitive quotations & meeting records
3.5.2.2.4 final selection
Dfd #: 3.5.2.2
Dfd Name: s/c select proc
Author: NF-SL
Date:
subcontract quotation

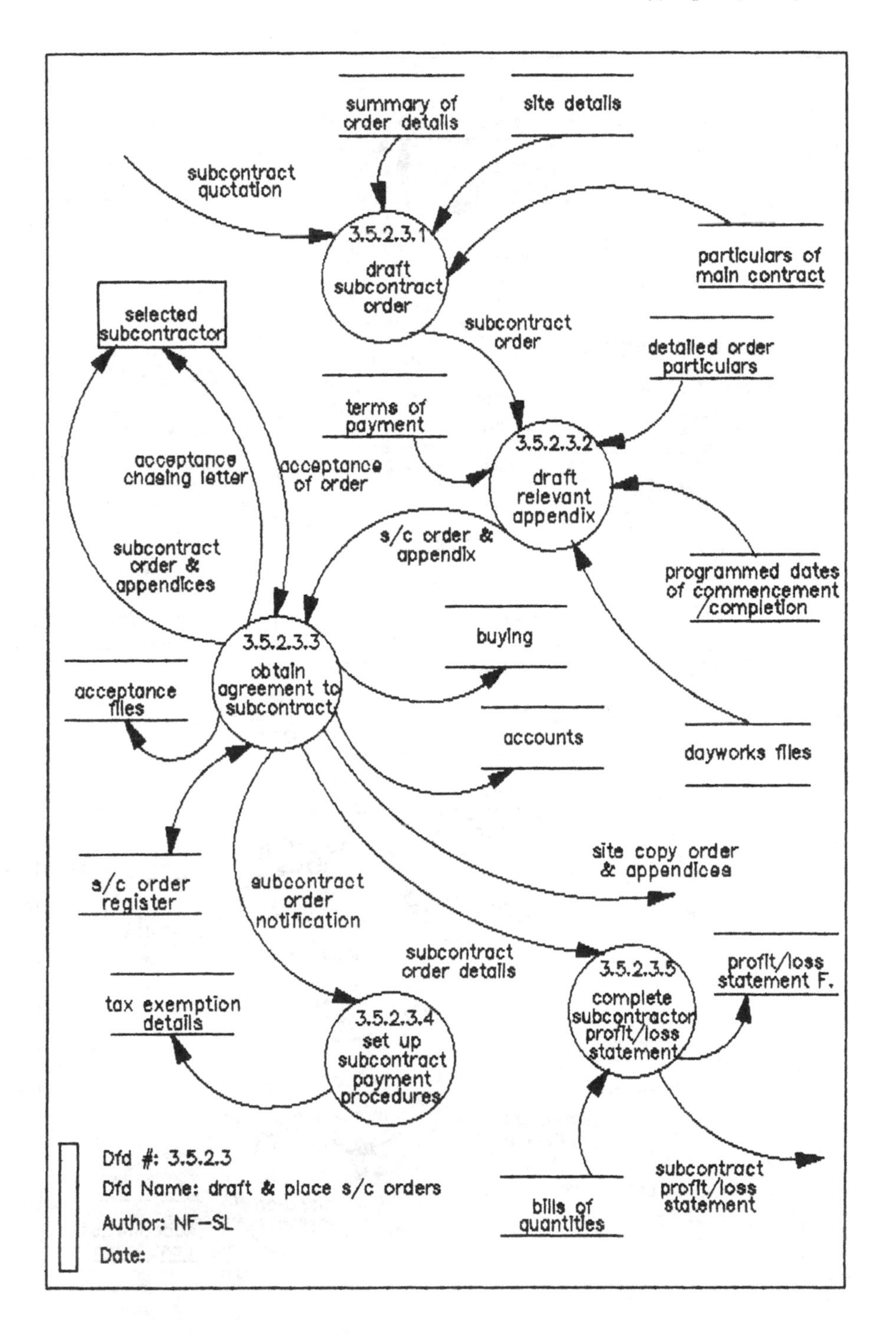
summary of order details
site details
subcontract quotation
3.5.2.3.1 draft subcontract order
particulars of main contract
selected subcontractor
subcontract order
detailed order particulars
terms of payment
3.5.2.3.2 draft relevant appendix
acceptance chasing letter
acceptance of order
s/c order & appendix
subcontract order & appendices
programmed dates of commencement /completion
buying
3.5.2.3.3 obtain agreement to subcontract
acceptance files
accounts
dayworks files
site copy order & appendices
s/c order register
subcontract order notification
subcontract order details
3.5.2.3.5 complete subcontractor profit/loss statement
profit/loss statement F.
tax exemption details
3.5.2.3.4 set up subcontract payment procedures
Dfd #: 3.5.2.3
Dfd Name: draft & place s/c orders
Author: NF-SL
Date:
subcontract profit/loss statement
bills of quantities

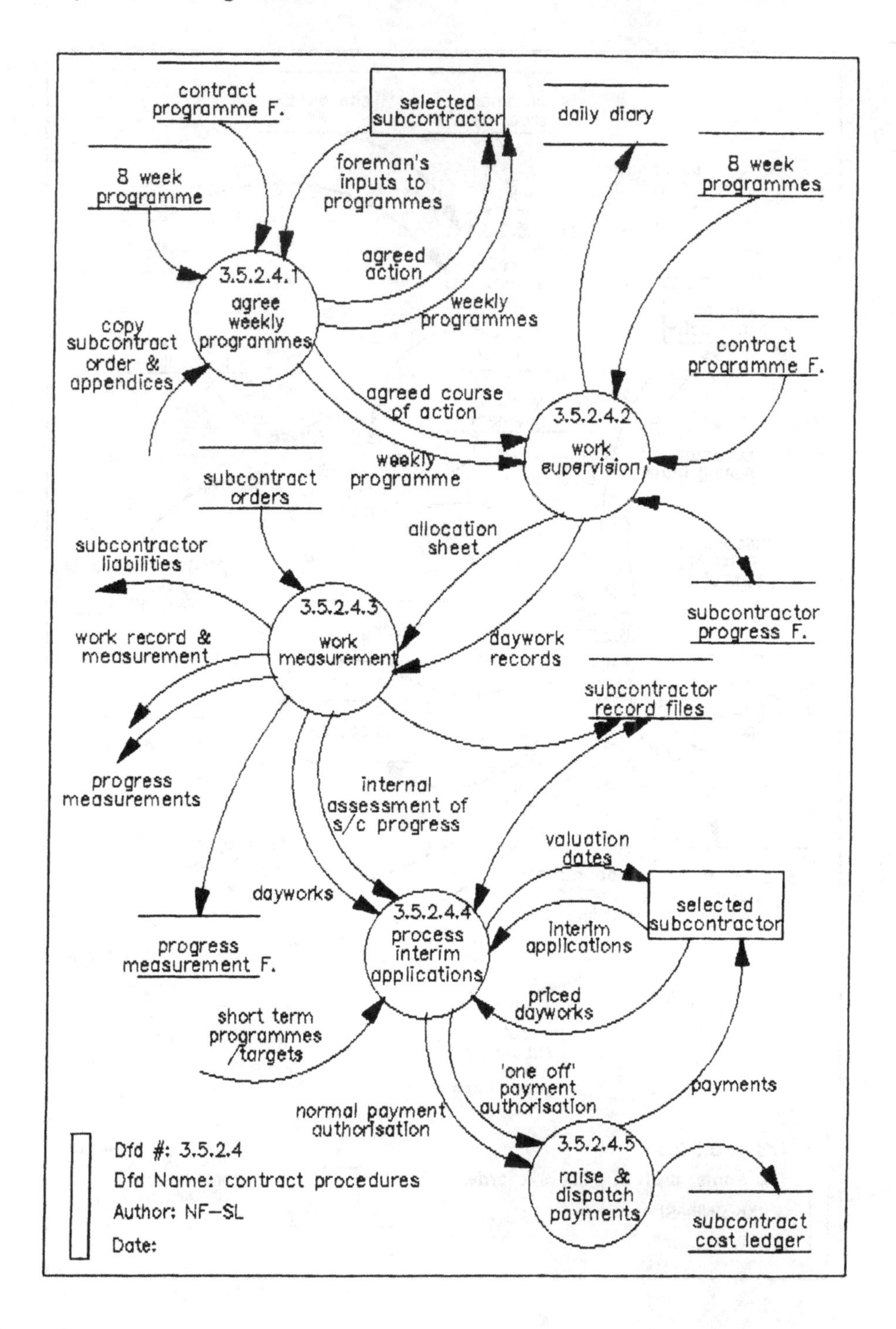
contract programme F.
selected subcontractor
daily diary
8 week programme
foreman's inputs to programmes
8 week programmes
agreed action
3.5.2.4.1 agree weekly programmes
weekly programmes
copy subcontract order & appendices
contract programme F.
agreed course of action
3.5.2.4.2 work supervision
weekly programme
subcontract orders
allocation sheet
subcontractor liabilities
3.5.2.4.3 work measurement
subcontractor progress F.
work record & measurement
daywork records
subcontractor record files
progress measurements
internal assessment of s/c progress
valuation dates
selected subcontractor
dayworks
3.5.2.4.4 process interim applications
interim applications
progress measurement F.
priced dayworks
short term programmes /targets
'one off' payment authorisation
payments
normal payment authorisation
3.5.2.4.5 raise & dispatch payments
subcontract cost ledger
Dfd #: 3.5.2.4
Dfd Name: contract procedures
Author: NF-SL
Date:

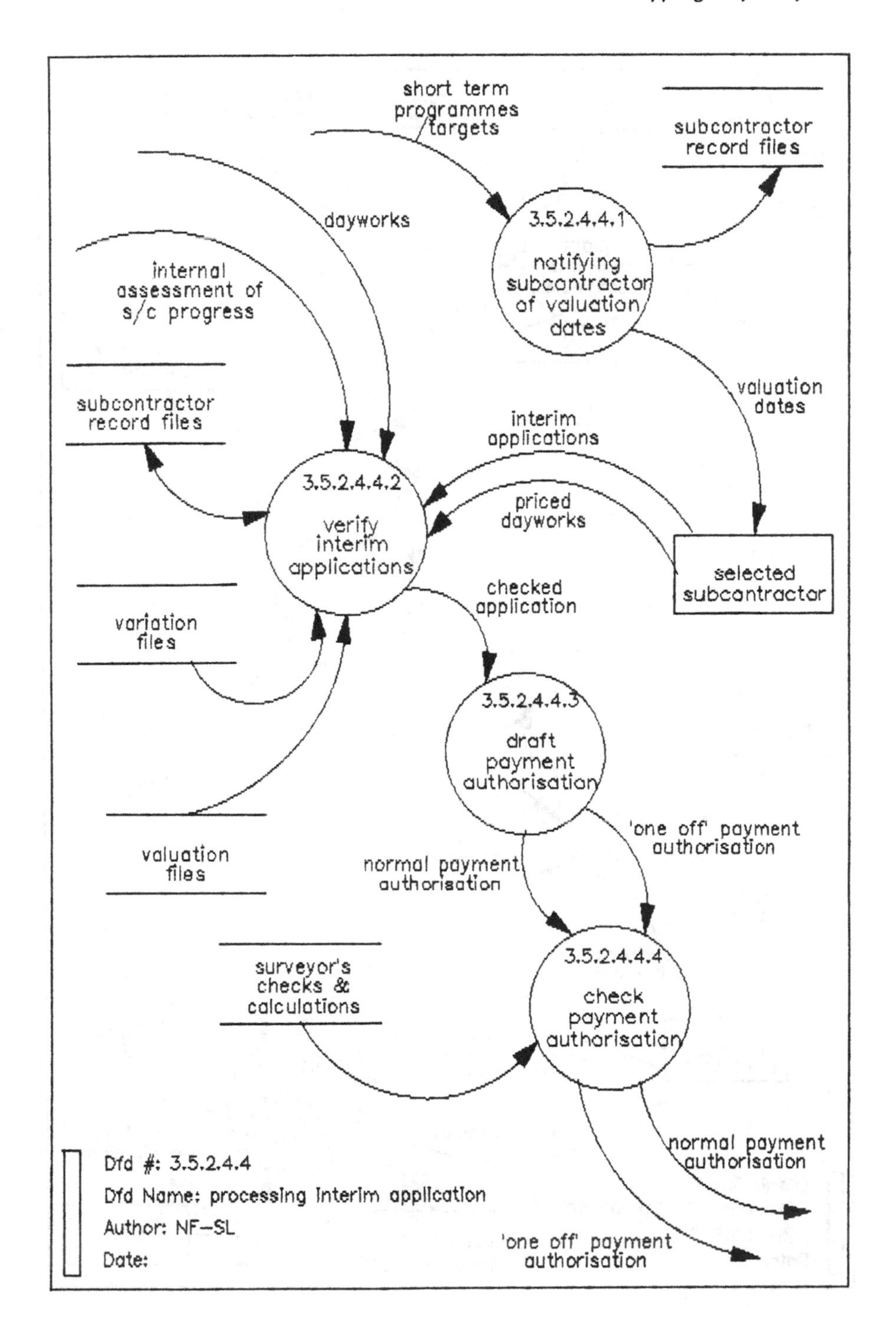
short term
programmes
targets
subcontractor
record files
dayworks
3.5.2.4.4.1
notifying
subcontractor
of valuation
dates
internal
assessment of
s/c progress
subcontractor
record files
valuation
dates
interim
applications
3.5.2.4.4.2
verify
interim
applications
priced
dayworks
selected
subcontractor
checked
application
variation
files
3.5.2.4.4.3
draft
payment
authorisation
valuation
files
'one off' payment
authorisation
normal payment
authorisation
surveyor's
checks &
calculations
3.5.2.4.4.4
check
payment
authorisation
normal payment
authorisation
Dfd #: 3.5.2.4.4
Dfd Name: processing interim application
Author: NF-SL
Date:
'one off' payment
authorisation

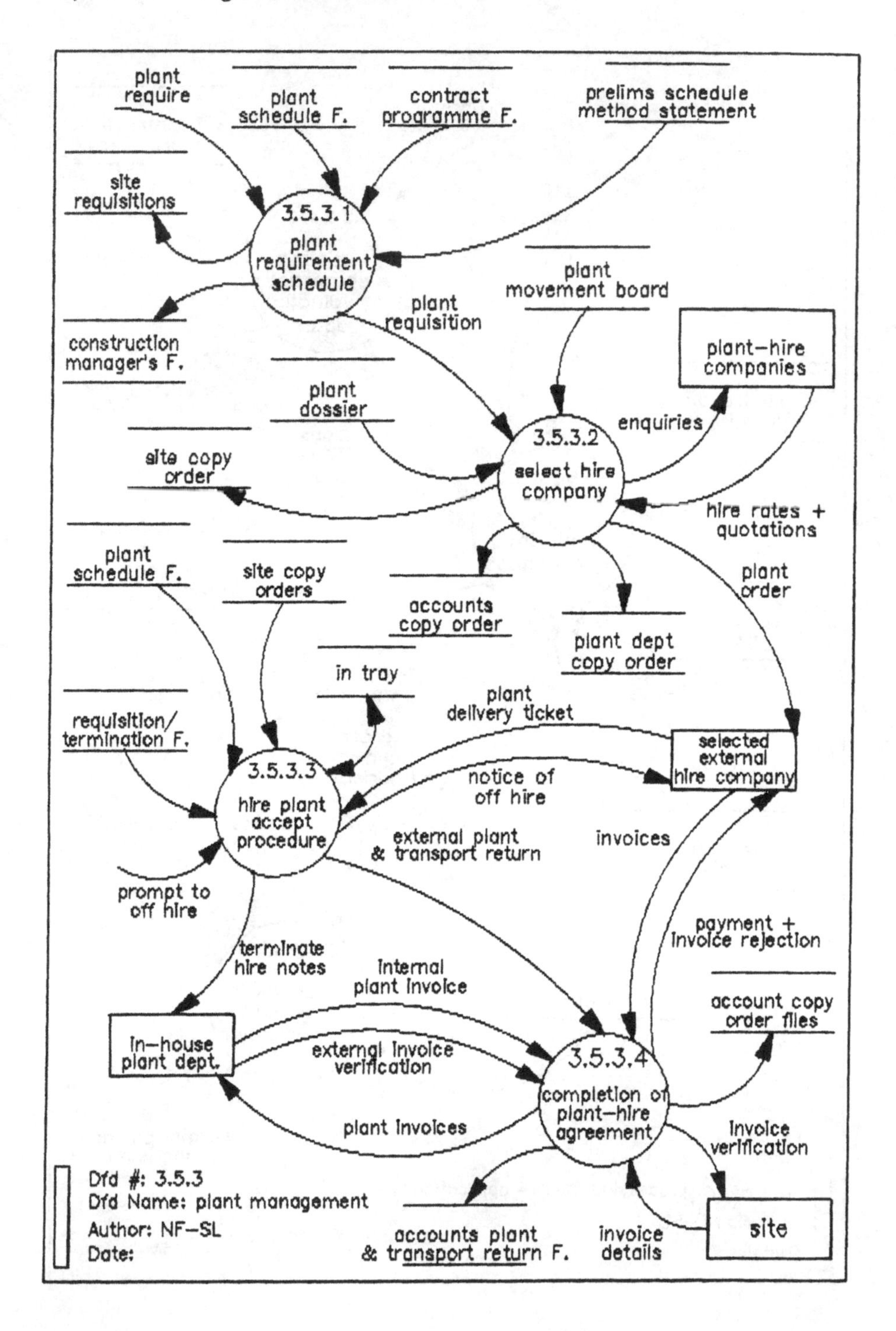
plant require
plant schedule F.
contract programme F.
prelims schedule method statement
site requisitions
3.5.3.1 plant requirement schedule
construction manager's F.
plant requisition
plant movement board
plant-hire companies
plant dossier
enquiries
3.5.3.2 select hire company
site copy order
hire rates + quotations
plant schedule F.
site copy orders
accounts copy order
plant dept copy order
plant order
in tray
plant delivery ticket
requisition/ termination F.
3.5.3.3 hire plant accept procedure
notice of off hire
selected external hire company
external plant & transport return
invoices
prompt to off hire
payment + invoice rejection
terminate hire notes
internal plant invoice
account copy order files
in-house plant dept.
external invoice verification
3.5.3.4 completion of plant-hire agreement
plant invoices
invoice verification
Dfd #: 3.5.3
Dfd Name: plant management
Author: NF-SL
Date:
accounts plant & transport return F.
invoice details
site

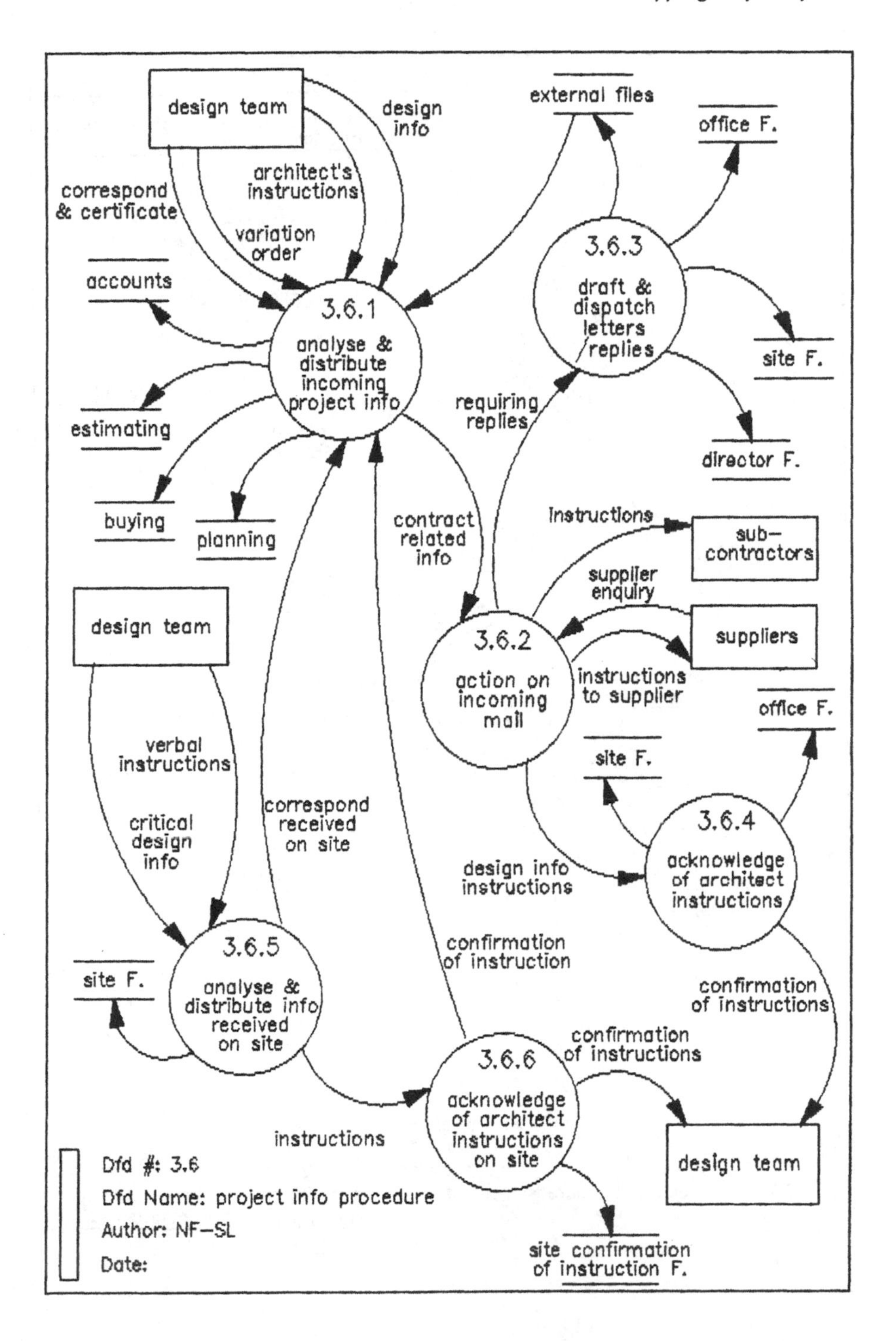
design team
design info
architect's instructions
correspond & certificate
variation order
external files
office F.
3.6.3
draft & dispatch letters /replies
accounts
3.6.1
analyse & distribute incoming project info
site F.
estimating
requiring replies
director F.
buying
planning
contract related info
instructions
sub-contractors
supplier enquiry
design team
suppliers
3.6.2
action on incoming mail
instructions to supplier
office F.
site F.
verbal instructions
critical design info
correspond received on site
3.6.4
acknowledge of architect instructions
design info instructions
confirmation of instruction
3.6.5
analyse & distribute info received on site
site F.
confirmation of instructions
confirmation of instructions
3.6.6
acknowledge of architect instructions on site
instructions
design team
Dfd #: 3.6
Dfd Name: project info procedure
Author: NF–SL
Date:
site confirmation of instruction F.

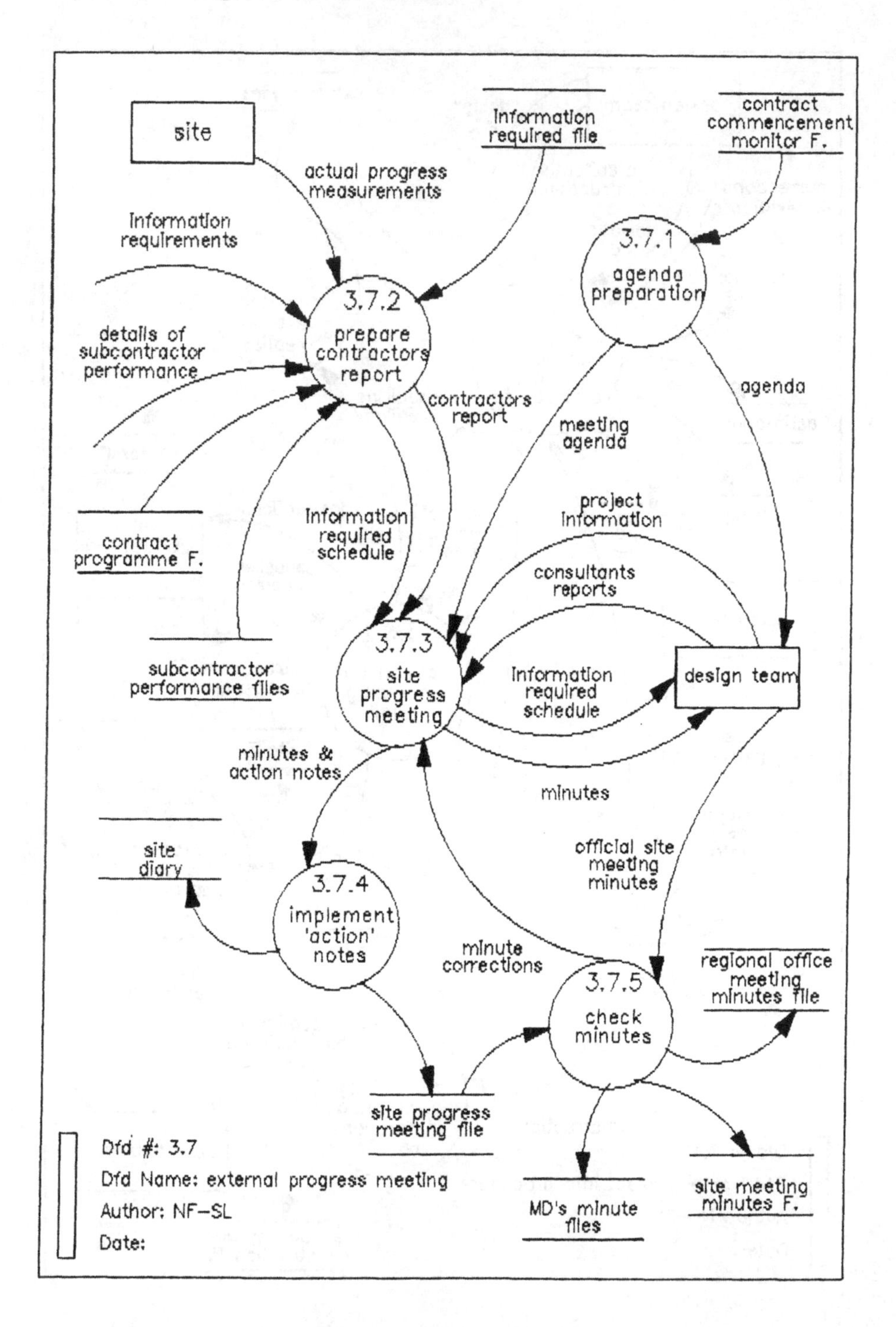
site
actual progress measurements
Information required file
contract commencement monitor F.
Information requirements
3.7.1 agenda preparation
3.7.2 prepare contractors report
details of subcontractor performance
contractors report
agenda
meeting agenda
project information
contract programme F.
Information required schedule
consultants reports
subcontractor performance files
3.7.3 site progress meeting
Information required schedule
design team
minutes & action notes
minutes
site diary
official site meeting minutes
3.7.4 implement 'action' notes
minute corrections
3.7.5 check minutes
regional office meeting minutes file
site progress meeting file
Dfd #: 3.7
Dfd Name: external progress meeting
Author: NF-SL
Date:
MD's minute files
site meeting minutes F.

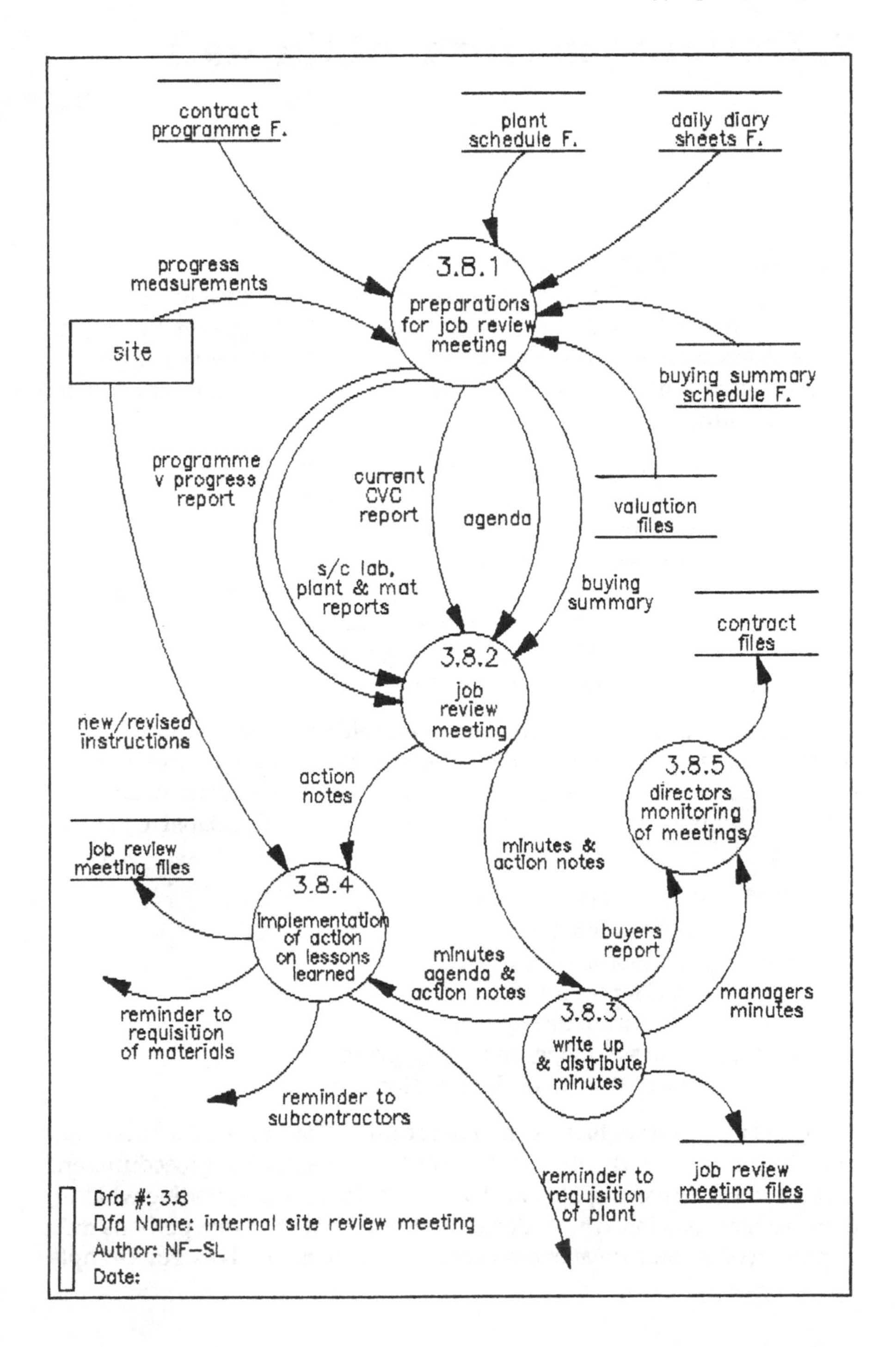
contract programme F.
plant schedule F.
daily diary sheets F.
progress measurements
site
3.8.1
preparations for job review meeting
buying summary schedule F.
programme v progress report
current CVC report
agenda
valuation files
s/c lab, plant & mat reports
buying summary
contract files
3.8.2
job review meeting
new/revised instructions
action notes
3.8.5
directors monitoring of meetings
job review meeting files
minutes & action notes
3.8.4
implementation of action on lessons learned
buyers report
minutes agenda & action notes
managers minutes
3.8.3
write up & distribute minutes
reminder to requisition of materials
reminder to subcontractors
reminder to requisition of plant
job review meeting files
Dfd #: 3.8
Dfd Name: internal site review meeting
Author: NF–SL
Date:

3. Defining the form of the data

Introduction to the data dictionary

This chapter presents the data dictionary. It lists and describes in alphabetical order all the data items used in the DFDs in chapter 2. It represents a complete list of data associated with the management of construction activity in a standard contracting company. All observed data have been classified into one of five groups:

(a) data element (e.g. invitation to tender, DFD #1)
(b) data group (e.g. info for head office, DFD #0)
(c) local file (e.g. tender files, DFD #1.2)
(d) process (e.g. decision to tender, DFD #1)
(e) terminator (e.g. design team, DFD #0). The DD is a set of definitions of all data flows, files, information source/sinks (terminators) and processes referred to in the DFDs. (Appendix 1 gives further explanation of the nature, content and purpsoe of a DD.)

Each of the unique names given to a data element displayed in a DFD is listed. For example, the system of (internal) project cost control as in the GDFM described in this book, centres around cost/value comparison (CVC) activity. There are seven entries in the DD that are directly related to the CVC process:

(a) CVC (data element)
(b) CVC procedures (process)
(c) CVC report (data element)
(d) Current CVC report (data element)
(e) CVC report - part (data element)
(f) CVC report to surveying director (process)
(g) CVC summary files (data element/local file).

Two of these entries have textual descriptions (mini-specs) which define what happens in a particular data element or process. CVC procedures and CVC report to surveying director have specifications that outline what the data element contains, what it does, what is done with it and, perhaps most important of all, the format or structure of the data themselves. For example, the process CVC procedures outlines the system for establishing accurate costs, profit and general financial performance on a project. It deals with

(a) the system for the preparation of accurate internal valuations
(b) the internal cost system for a project
(c) the transformation of site data into the structure used on a CVC form
(d) the internal managerial consultation process (mainly on reserves, both official and hidden), at the draft CVC report stage
(e) the CVC reporting system to directors
(f) the official CVC reporting information for the company accounting system.

In this GDFM model the level of detail offered has been deliberately controlled. For example, it would have been possible to show how contractors prepare and enter the appropriate data on to standard company forms. This approach has been deliberately resisted, as such detail would have masked the true purpose of the model - to show the principles, structure, logic or algorithm behind a contractor's project management system. The operational detail is the domain of the experienced up-to-date practising manager. This GDFM attempts to exemplify a typical structure to have behind the data. In a similar way, every human being is unique, but there are enough common principles known about how a human body and mind works to allow a medical expert to treat disease and malfunctions. (For a discussion on the level of underlying commonality between the systems of various contractors see Appendix 1.)

Communication between groups

An interface point exists at points of information exchange between different systems or sub-systems, usually within companies (between departments or subsidiaries) and between companies or groups. The information at an interface is exchanged across a context boundary between discrete sub-systems.

Most information problems occur at interfaces of one kind or another, particularly those of management information, where human nature comes so strongly into play. Again, if the data at the interface point are systemized, consistency and reliability will be increased and considerable improvements on information transfer achieved. As in other areas, if data which have to flow across an interface are systemized, then the system structure will be able to act, at the very least, as a checklist for the operational manager.

Data items and descriptions

AC'S VARIATION FILES
Entry Type : Data Element/Local File

ACCEPT INVITATION
Entry Type : Data Element

ACCEPTANCE CHASING LETTER
Entry Type : Data Element

ACCEPTANCE FILES
Entry Type : Data Element/Local File

ACCEPTANCE OF ORDER
Entry Type : Data Element

ACCEPTING OF ORDER
Entry Type : Data Element

The contractor will accept usually the lowest tender by a group of specialist sub-contractors. However, other factors will affect the decision on which to accept. Factors such as:

Terms and conditions that the sub-contractor is trying to impose, exclusions such as provision of temporary works, access and other attendances.

ACCOUNT COPY ORDER FILES
Entry Type : Data Element/Local File

ACCOUNTS
Entry Type : Data Element/Local File

ACCOUNTS CONTRACT F.
Entry Type : Data Element/Local File

ACCOUNTS COPY ORDER
Entry Type : Data Element/Local File

ACCOUNTS F.
Entry Type : Data Element/Local File

ACCOUNTS PAYABLE SCHEDULE
Entry Type : Data Element/Local File

ACCOUNTS PLANT & TRANSPORTATION RETURN F.
Entry Type : Data Element/Local File

ACCRUALS
Entry Type : Data Element

ACCRUALS LABOUR COSTS
Entry Type : Data Element

ACKNOWLEDGEMENT OF ARCHITECT INSTRUCTIONS
Entry Type : Process

Specification:

This process deals with the following data : office F., site F., design info instruction, confirmation of instructions.

The Construction Manager is usually responsible for confirming the receipt of all Architects instructions and drawing issues. This he does on a Confirmation of Instructions Sheet which is hand written and dispatched to the Architect with copies being sent to the Consulting Engineer and the PQS. Copies of the Instructions and drawings are normally dispatched to site and a copy of each is also retained in the office. Copies of the Confirmation of Instruction Sheets are also filed both in the office and on site.

ACKNOWLEDGEMENT OF ARCHITECT INSTRUCTIONS ON SITE
Entry Type : Process

Specification:

This process deals with the following data: instructions, confirmation of instruction, confirmation of instructions, site confirmation of instruction F.

All verbal instructions received on site are normally confirmed by the agent on Confirmation of Instructions

Sheets. As with the office confirmations of written instructions, copies of the sheet are sent to the Architect, Consulting Engineer and PQS. Internal site and office copies are also kept. The office copy is dispatched via the Construction Manager to the Surveyor who files it in the office files.

ACTION NOTES
Entry Type : Data Element

ACTION ON INCOMING MAIL
Entry Type : Process

Specification:

This process deals with the following data: requiring replies, contract related info, instructions, supplier enquiry, instructions to supplier, design info instructions.

Mail relating to company matters is usually retained for filing by the directors whom together decide who will handle it. The Surveying director retains the original copies of architects certificates and presents them to the client straight away. Other certificates such as for extensions of time and practical completion are also normally retained and filed with the contract documents.

For promotion related correspondence the Construction Manager will normally sort out and reply to matters arising. Consultations with the agent over such items will normally consist of frequent telephone calls to site, most days, with visits at least one a week. All original correspondence is generally filed in the central office file looked after by the surveyor. Copies of production mail may be retailed by the Manager however, for further reference.

Architects instructions and design information issued and received in the mail are usually analysed by the Construction Manager. Subcontractors and nominated suppliers implicated by the instructions are normally dispatched a copy with a covering standard letter.

The Surveyor is usually responsible for sorting out financial problems on the contract but would normally be in close liaison with the Construction Manager and Surveying Director over the more important aspects.

ACTIVITY DATES
Entry Type : Data Element

ACTUAL PROGRESS MEASUREMENTS
Entry Type : Data Element

ADDITIONAL ENQUIRIES
Entry Type : Data Element

ADJUST COST & COMPLETE COST SCHEDULE
Entry Type : Process

Specification:

This process deals with the following data: surveyor approved costs, monthly cost and copy invoices, schedule of costs and adjustments.

Upon receipt of the months cost report a surveyor will normally transfer the cost centre total costs (cost to date plus accruals) to the corresponding column in the schedule of Costs and Adjustments. The cost totals should then be checked and where necessary adjustments are made to reveal the Adjusted Cost Total.

With his greater detailed knowledge of the contract the contractor's surveyor should briefly check the costing of the plant and materials to ensure that no significant items have been missed or wrongly allocated. This can be done by quick reference to copy invoices attached to the report and to the weekly transaction reports. Any obvious omissions or wrong allocations can be corrected. The costs of subcontractors and suppliers shown on the cost report have to be deducted from the surveyors assessed liabilities to reveal the 'Add', value necessary to adjust them to 'true costs'. (The costs on the cost report will only reflect the value of the last payment to pass through the Accounts department). Having checked all the costs the Adjusted Costs are added down the page to reveal the 'Total Adjusted Cost'.

ADJUSTED TRUE VALUE
Entry Type : Data Element

AGENDA
Entry Type : Data Element

AGENDA PREPARATION
Entry Type : Process

Specification:

This process deals with the following data: contract commencement monitor F, agenda, meeting agenda

A company will normally have a standard site progress meeting agenda which is issued on each contract as part of the contract commencement monitor. Prior to the first site progress meeting the Construction Manager distributes the agenda to the parties who will be in attendance. Many Architectural practises have developed their own agendas and hence prefer not to use the company's own agenda. The Construction Manager and Site Agent ensure that all matters listed on their agenda are raised and discussed at the meeting irrespective of which agenda is used.

AGREE WEEKLY PROGRAMMES
Entry Type : Process

Specification

This process deals with the following data : 8 week programme, contract programme F, foreman's inputs to programmes, agreed action, weekly programmes, agreed course of action, weekly programme, copy subcontract order & appendices.

Working typically from a Contract Programme and the eight week short term programme, the Site Agent and General Foreman meet together at the end of each week to plan in detail the following weeks work. The Foreman and 'gangers' representing the principal subcontractors are consulted during the preparation of these programmes to ensure their full agreement and commitment to them. A copy of the weekly programme is subsequently given to each foreman. The Site Agent monitors the actual performance during the week to ensure the programmed work is completed.

AGREED ACTION
Entry Type : Data Element

This deals with agreements on short term working between the Site Agent and the subcontractors.

AGREED COURSE OF ACTION
Entry Type : Data Element

AGREED HIRE RATES
Entry Type : Data Element

AGREED WEEKLY PROGRAMMES

Entry Type : Data Element

AGREEMENT ON 1st MEETING
Entry Type : Data Element

Date and agenda for a first meeting between the contractor and the client and his advisers is agreed.

ALL CURRENT CONTRACT CVC REPORTS
Entry Type : Data Element

ALL IN RATES & QUOTATIONS
Entry Type : Data Element

ALLOCATION SHEET
Entry Type : Data Element

ALLOCATION SHEETS
Entry Type : Data Element/Local File

ANALYSE & DISTRIBUTE INCOMING PROJECT INFO
Entry Type : Process

Specification:

This process deals with the following data: external files, design info, architect's instructions, variation order, correspond & certificate, accounts, estimating, buying, planning, correspond received on site, confirmation of instruction, contract related info.

A company normally requests that all typical correspondence and design information is sent direct to the Regional Office and not Site. The exception to this would typically be a very large contract with a site based Construction manager which would be run from site. Upon receipt of post, letters of major importance; certificates from the Architect, letters of dispute from subcontractors, etc., would be retained by a director for further consideration later in the morning. Mail for the service departments, accounts, buying, estimating and planning is passed to the appropriate departmental head for distribution and processing. Information relating to current contracts is sorted between that which is for the Construction Manager (production related) and that which is for the Surveyor (financially related). All production related mail is copied to site and then passed to thee Manager. The Surveying Director should he wish to know more about any letters, marks it up with applicable questions or queries before distribution. If he considers that the Managing Director or he should receive a copy of any mail it is again marked up appropriately.

ANALYSE & DISTRIBUTE INFO RECEIVED ON SITE
Entry Type : Process

Specification:

This process deals with the following data: correspond received on site, verbal instructions, critical design info, site F., instructions.

In theory all that should be received on site are verbal instructions from the design team or clerk of works and copy correspondence from the Regional office. However, there is usually an amount of misdirected mail and also late or critical design information received as well.

Verbal instructions are normally confirmed by the Agent direct from site (see section 3.6.6 DFD No.3.6). Critical design information issued to site should as requested (at the beginning of the contract) be copied to the regional office (with the office copies from where it will generally be confirmed by the construction manager.

Misdirected mail is forwarded direct to regional office for copying and distribution or if urgent is retained by the agent who consults the construction manager by phone to agree a course of action. Copy correspondence from Regional Office is noted and filed as appropriate.

ANALYSED VALUATIONS
Entry Type : Data Element

ANALYTICAL B.Q.
Entry Type : Data Element

ANALYTICAL BILLS OF QUANTITIES
Entry Type : Data Element/Local File

ARCHITECT CERTIFICATE
Entry Type : Data Element

A copy of the official document passed to the client for payment.

ARCHITECT PQS & CLIENT FILES
Entry Type : Data Element

ARCHITECT'S INSTRUCTIONS
Entry Type : Data Element

ARCHITECTS CERTIFICATE FILES
Entry Type : Data Element/Local File

AREA MANAGER COMMENTS RE PROF & PRELIM SCHED
Entry Type : Data Element

ARITHMETICAL CHECKING
Entry Type : Process

Specification:

This process deals with the following data: built up & extensions & work sheets, build up & extensions, verified build ups & extensions, build ups extensions & work sheets.

The Estimating Department's comptometer operators check all worksheets, extensions, measured rates, page and section/trade totals. Mistakes are corrected when applicable and all information is passed back to the Estimator as soon as possible. The Estimator briefly checks any corrected items to ensure all consequential mistakes have also been rectified.

ARITHMETICALLY CORRECT INVOICES
Entry Type : Data Element

ARRANGE DELIVERY
Entry Type : Process

Specification:

This process deals with the following data: site order files, instructions to supplier files, call off request, confirmation of call off, site F, regional office file, buyer, accounts, materials shortage.

For 'one off' specific orders it is usual not to have to call off the delivery as the supplier will have been given the due date on the order. For bulk orders however when more than one delivery is usually required the Agent uses a Confirmation of Call Off Instruction/Request to supply sheet to communicate the required delivery date and quantity of the materials concerned.

Materials are usually called-off by telephone and subsequently confirmed using the above form. The Agent is required to enter full details of the materials required and to quote the order number covering the supply.

The sheet is also used by site as a local purchase order but is only usable at the merchants with whom a covering general order has been placed. In such a case the form will be headed 'Request to Supply' and will schedule the goods required. The buyer is able to monitor the goods purchased in this way through the file copy he receives and ensures the system is not abused.

ASSESS PRE-TENDER QUOTATIONS
Entry Type : Process

Specification:

This process will deal with the following data: defined requirement, tender enquiry abstracts F, pretender quotations, supplier directory, competitive pretender quotations, additional enquiries.

Having defined and specified the precise requirements a buyer should examine the relevant quotations received at the pre-tender stage. The tender enquiry abstracts should detail the suppliers to whom enquiries were originally sent and the quotations and quotation analysis sheets (for the larger supply items only) reflect the suppliers whom actually submitted quotations. The buyer should decide whether additional quotations are necessary and if so dispatch new enquiries. For 'materials required due to a variation in the contract works' no pre-tender quotations will be on hand and several new enquiries will be required. Enquiries should contain all applicable extracts from the bills of quantities and specification, or in the case of variations a full description of the components. If little time is available quotations can be obtained direct from the supplier by telephone.

The buyer may wish to re-examine the quotations received at the pre-tender stage for restrictive terms and conditions and for suitability in light of fully specified requirements. The competitive quotations are retained for comparison against any new quotations.

ASSESS THE LEVEL OF FLUCTUATIONS
Entry Type : Process

Specification:

This process deals with the following data: labour records, material increase file, fluctuation totals

The contract conditions will normally determine whether it is a fixed price or fluctuating contract. If fluctuating, several choices of provisions exist which include:

1. contribution, levy and tax fluctuations
2. labour and materials cost and tax fluctuations
3. price adjustment formulae (NEDO formula).

ASSESS VALUE OF MATERIALS ON SITE
Entry Type : Process

Specification:

This process deals with the following data: total materials on site, price books, material orders, materials on site.

The materials on site have not yet been incorporated into the building are normally included in the valuation and valued at cost price to the contractor. An inspection of the site is generally made by the site Surveyor as near to the valuation date as possible and all materials and quantities noted. The materials have to be listed, specified and prices. This will necessitate reference to the materials orders, records of materials delivered and price books/lists. It is frequently the job of a junior or assistant site Surveyor to prepare this section of the valuation.

Materials ordered and pair for but which are not on site can be incorporated into the valuation provided that proof of this can be shown and if they are clearly marked and set aside in the suppliers / subcontractors yard. It is the task of the job Surveyor to make the appropriate visit(s) to suppliers and subcontractors yards to obtain evidence that these requirements have been met.

ASSESS VALUE OF VARIATION
Entry Type : Process

Specification:

This process deals with the following data: measurement of variations, ac's variation files, confirmation of instructions files, measurement files, plant return orders, allocation sheets, materials returns orders, variations total.

Specified changes to the original content of work are included in the valuation under the general heading of variations. Architects/Engineers instructions, verbal instructions and changes indicated on revised drawings are acknowledged by the company on 'Confirmation of Instruction Sheets'. Quotations for the work involved with these variations may have been submitted, in which case they would form the basis of the value to be included in the account. For items not covered by a quotation, however, the value may need to be calculated from first principles, reference to their labour plant orders would be used to gather supporting information for the pricing of such variations.

ASSESS VALUES OF DAYWORKS
Entry Type : Process

Specification:

This process deals with the following data : total value of dayworks, materials returns orders, allocation sheets, plant returns orders, dayworks files, foreman/agent's records.

On occasions when work is undertaken which is not part of the original contract; and is not covered by a variation order, or when it is the result of a variation but does not warrant measuring in the traditional manner; it is recorded on a dayworks sheets. This is verified by the Architect/Clerk of Works and submitted to the PQS prices in accordance with predetermined labour rates and applicable plant, material and transport rates.

A summary of the dayworks is prepared for inclusion in the valuation and the section total carried forward to the summary.

Daywork sheets not yet verified, priced or agreed with the relevant consultants will be priced and included in the valuation anyway. Payment of such items will be at the discretion of the PQS.

AUTHORISE AND DISPATCH CHEQUE
Entry Type : Process

Specification:

This process deals with the following data: invoice payment rejection, remittance advice and cheque.

Normally a cheque issuing system requires two signatures. Normally one would be an accountant and the second a director.

AUTHORISED DAYWORK
Entry Type : Data Element

AUTHORISED DAYWORKS
Entry Type : Data Element

BILLS OF QUANTITIES
Entry Type : Data Element/Local File

BILLS OF QUANTITIES SPEC. DRAWINGS
Entry Type : Data Element/Local File

BILLS OF QUANTITY DETAILS
Entry Type : Data Element/Local File

BONUSES
Entry Type : Data Element

Incentive payment system for directly employed staff.

BOOK COST & CASH RECONCILIATION
Entry Type : Data Element/Local File

BOQS
Entry Type : Data Element

The priced Bill of Quantities is submitted to the client by the contractor.

BOUGHT LEDGER
Entry Type : Data Element/Local File

BUILD UP & EXTENSIONS
Entry Type : Data Element

BUILD UP & EXTENSIONS & WORK SHEET
Entry Type : Data Element

BUILD UP COMPLETE ESTIMATE
Entry Type : Process

A complete estimate is prepared that represents the total cost of all construction work, including preliminaries outlined in the tender documentation.

BUILD UP UNIT RATES
Entry Type : Process

Specification:

This process will deal with the following data : all in rates & quotations, established unit rates and quotations.

As mentioned in section 1.2.6 little work is directly priced by the companies own Estimators. For the trades that are sometimes priced, groundwork, carpentry and brickwork, unit rates have to be built up for each measured item in the bill of quantities. Unit rates usually consist of elements for labour, plant and materials. The cost of each element is not recorded separately although pricing notes are kept to illustrate how the unit rates have been built up.

The subcontract tenders selected in 1.2.6 are re-examined and checked to ensure allowances have been included for items such as unloading, storage and protection of materials, provision of plant, offices, scaffolding, water, electricity and telephone and protection of finished work. Sums must be added to the tenders to cover those attendances which involve a cost.

BUILD UPS EXTENSIONS & WORK SHEETS
Entry Type : Data Element

BULK MATERIALS RECONCILIATION FILE
Entry Type : Data Element/Local File

BULK QUANTITIES
Entry Type : Data Element

BUYER
Entry Type : Data Element/Local File

BUYER & REGIONAL OFFICE COPY F
Entry Type : Data Element/Local File

BUYER'S INPUTS
Entry Type : Data Element

BUYERS REPORT
Entry Type : Data Element

BUYING
Entry Type : Data Element/Local File

BUYING FILE (COPY ORDER)
Entry Type : Process

Specification:

This process deals with the following data: tender enquiry abstracts, subcontract quotations, s/c quotes comments, selected quotations, revised quotations, negotiation and negotiated savings and expected margins on S/Cs.

Upon receipt of a batch of Tender Enquiry Abstracts the Chief Buyer seeks to add the names and addresses of additional sub-contractors and suppliers who would submit keen prices. Typically a buying department has its own directly of subcontractors arranged by trade, which simply lists the names and addresses of companies known to work in each field of operation. It has a more comprehensive directory of material suppliers.

The Buyer concentrates primarily on the subcontract abstracts and usually lists an additional two or three companies on each. Bulk material abstracts may also be considered in the same manner. The Contracts Manager associated with the tender may also be consulted as to possible further companies he would like to see included on the abstract. The abstract sheets are passed straight back to the enquiry clerk for the additional invitations to tender to be drafted.

BUYING SUMMARY
Entry Type : Data Element

BUYING SUMMARY SCHEDULE
Entry Type : Data Element. See Figure 4 (a-k)

BUYING SUMMARY SCHEDULE F
Entry Type : Data Element/Local File. See Figure 4 (a-k)

BUYING SUMMARY SCHEDULE (FROM 3.1)

Buying summary schedule
of Covering Orders for Sundries/Consumables **Contract** ....

General light side materials (Local Merchant)	
Sawn & prepared timber (small quants only)	
Bar spacers	
Nails & woodscrews	
Plastic plugs & fixings	
Abrasive discs & drill bits	
Shot-fixing nails & cartridges/ free loan cartridge tools	
Expansion joint material (i.e. fiberpak/flexcell)	
Polystyrene	
Gas oil/Diesel/Engine oil	
Mould oil	
Refills only, for First Aid Cabinets	
Roller Towel or Paper Towel service	
Concrete cube-testing service	
Calor Gas supply in Bulk system (Cylinders are Plant Div' responsibility)	

1 SHOULD INDICATE WITH A TICK THOSE ITEMS FOR WHICH AN ORDER IS REQUIRED. REQUISITION IS NOT NECESSARY AS THESE ITEMS ARE SUBJECT TO ANNUAL RATES AGREED BY BUYING DEPT.

2 SITE SUPERVISOR SHOULD CALL REQUIREMENTS FORWARD DIRECT FROM SUPPLIER QUOTING THE BUYER'S ORDER NO. IN FULL & CONFIRMING CALL OFF IN WRITING

3 MATERIALS & SERVICES LISTED HERE SHOULD NOT BE CALLED FORWARD FROM SUPPLIERS OTHER THAN THOSE INDICATED ON THE OFFICIAL ORDERS

NB: THIS SHEET IS INTENDED TO SERVE AS REQUISITION FOR BUYER

Fig. D4(a). Buying summary schedule : sundries

Buying summary schedule
of Covering Orders for Bulk Materials/General **Contract** ....

Description of Materials	1st Load req'd Approx	BQ Ref & Spec	
O.P. Cement/Masonry Cement			1 SITE SUPERVISOR MUST REQUISTION THESE MATERIALS STATING TOTAL QUANTITIES AND DELIVERY PROGRAMME
Steel Rod Reinforcement			
Fabric Reinforcement			
Polythene Film			2 BUYER WILL PLACE A COVERING ORDER
'Hepworth' Drainage			
Pitch Fibre Drainage			3 SITE SUPERVISOR SHOULD CALL REQUIREMENTS FORWARD DIRECT FROM SUPPLIER QUOTING BUYER'S ORDER NO. IN FULL & CONFIRMING CALL OFF IN WRITING
Ready-mixed Mortars			
Carcassing Softwood			
Concrete Chamber Rings			
Heavy-side Materials - miscellaneous			

NB: THIS SHEET IS INTENDED TO SERVE AS REQUISITION FOR BUYER

Fig. D4(b). Buying summary schedule: general bulk materials

Buying summary schedule Hardcore/Aggregates
of Covering Orders for Bulk Materials/

Contract..

Type & Grade of Materials	Quantity		1st Load req'd Approx	BQ Ref & Spec
40mm (11/2") Graded Shingle		Tonnes		
20mm (3/4") Graded Shingle		Tonnes		
10mm (3/8") Graded Shingle		Tonnes		
6mm (1/4") Graded Shingle		Tonnes		
40mm (11/2") All in Ballast		Tonnes		
20mm (3/4") All in Ballast		Tonnes		
10mm (3/8") All in Ballast		Tonnes		
Sharp washed sand		Cu. Yds		
Building sand		Cu. Yds		
Hardcore		Cu. Yds		
Ashes/Clinker		Cu. Yds		
Rejects		Tonnes		
Hoggin		Tonnes		
Granular fill		Tonnes		
MOT type 2 fill		Cu. Yds		
Muck away-Hand Load		Cu. Yds		
Muck away-Machine Load		Cu. Yds		

1
CONVERSION FACTORS ARE:
1 TON = 1.016 TONNES
$1M^3$ = 1.308 YDS^3 (APPROX)
27 CWTS = $1YD^3$

2
THE SITE SUPERVISOR/ MANAGER SHOULD PROVIDE ACCURATE QUANTITIES ON THIS SHEET

NB: THIS SHEET IS INTENDED TO SERVE AS REQUISITION FOR BUYER

Fig. D4(c). Buying summary schedule: hardcore aggregates

Buying summary schedule of Covering Orders for Bulk Materials/ Ready Mix Concrete

Contract ....

Quantity of each mix	1st Load reqd approx	Mix Details	B.Q. Reference & Spec.

1. PLEASE ENSURE ALL MIXES ARE INCLUDED E.G. TEMP. WORKS etc./PUMP MIXES etc.
2. DELIVERY DATES SHOULD INCLUDE DETAILS OF ANY LARGE POURS THAT ARE REQUIRED
3. THE USE OF ANY ADDITIVES IS SUBJECT TO THE PRIOR APPROVAL OF OUR CONSULTANT ENGINEERS; AND SHOULD NOT BE CALLED FORWARD BY SITE

NB: THIS SHEET IS INTENDED TO SERVE AS REQUISITION FOR BUYER

Fig. 4(d). Buying summary schedule: ready mix concrete

Buying summary schedule - Bricks -
of Covering Orders for Bulk Materials/

Contract ....

Quantity	Type of Brick	Subject of Reservation (Yes/No)	Delivery Programme & whether packaging/palleting & mech off-loading reqd.	B.Q. Reference & Spec

NB: THIS SHEET IS INTENDED TO SERVE AS REQUISITION FOR BUYER

Fig. 4(e). Buying summary schedule: bricks

Buying summary schedule - Blocks -
of Covering Orders for Bulk Materials/ **Contract** ....

Thickness (in mm)	Quanity (in M^2)	Standard or Fairfaced	Type of Block (name & whether hollow/solid etc.)	Delivery Programme & whether packaging/palleting & mech off-loading reqd.	B.Q. Reference & Spec.

NB: THIS SHEET IS INTENDED TO SERVE AS REQUISITION FOR BUYER

Fig. D4(f). Buying summary schedule: blocks

Buying summary schedule
Materials (Miscellaneous)

Contract ....

Schedule Ref No.	Details to be provided by, or Bulk Order	Latest Requisition Date	Required on Site by	DESCRIPTION OF MATERIAL	B.Q. REFERENCE	Order No and Date	NAME OF SUPPLIER OR SUBCONTRACTOR

THE SITE SUPERVISOR MUST RAISE REQUISITIONS FOR ALL ITEMS ABOVE

Fig. D4(g). Buying summary schedule: miscellaneous

Buying summary schedule
PC Sums - Nominated Supplier

Contract ....

Schedule Ref No.	Details to be provided	Latest Requisition Date	Required on Site by	DESCRIPTION OF MATERIAL	B.Q. REFERENCE	Order No and Date	NAME OF SUPPLIER

Fig. D4(h). Buying summary schedule: nominated suppliers

Bulk summary schedule
Provisional Sums

Contract ....

Schedule Ref No.	Details to be provided by	Latest Requisition Date	Required on Site by	DESCRIPTION OF MATERIALS OR WORKS	B.Q. REFERENCE	Order No and Date	NAME OF SUPPLIER OR SUBCONTRACTOR

Fig. D4(i). Buying summary schedule: provisional sums

Buying summary schedule
Direct Sub-Contractors

Contract ....

Schedule Ref No.	Details to be provided by	Latest Requisition Date	Required on Site by	DESCRIPTION OF TRADE	B.Q. REFERENCE	Order No and Date	NAME OF SUB-CONTRACTOR

Fig. D4(j). Buying summary schedule: direct subcontractors

Buying summary schedule
Labour only Sub-Contractors

Contract ....

Schedule Ref No.	Details to be provided by	Latest Requisition Date	Required on Site by	DESCRIPTION OF TRADE	B.Q. REFERENCE	Order No and Date	NAME OF SUB-CONTRACTOR

Fig. D4(k). Buying summary schedule: labour-only subcontractors

Entry Type : Data Group

Composition:
BUYING SUMMARY SCHEDULE + BUYING SUMMARY SCHEDULE (FROM 3.1)

BUYING SUMMARY SCHEDULE (TO 3.5)
Entry Type : Data Element

CALCULATE ACCRUALS
Entry Type : Process

Specification:

This process deals with the following data: accruals, labour costs, invoices received but not costed, material returns, plant returns, accruals.

Having produced the most up to date costs possible the costing personnel should calculate the sum of the costs which have been incurred but not yet costed. These costs are termed accruals and are calculated in two different ways. Firstly the costs of direct labour, labour expenses, and labour bonuses have to be assessed for the latter weeks of the month. This can be achieved by examining the last known actual weekly costs and assuming that the labour resources on site have not changed projecting those costs forward for each week missing. The actual labour resources can be checked against the site time sheets and the supervisors Monthly Movement and claim forms providing both have been received.

A second method of calculating the accrual costs is used to assess the known outstanding liabilities in respect of plant and materials. At the time of calculating the costs some invoices have probably not been received from suppliers although the goods have been delivered to site. An accounts department should monitor which goods have been invoiced by crossing through appropriate entries on the material return sheets/plant and transport returns, as invoices are received. At the time of preparing the costs, entries on the sheets not crossed represent the items for which costs need to be calculated. This costing is done with the copy material/plant orders revealing the appropriate rates. The material and plant returns each have columns for entering the 'Forecast Cost' which represents these accrued costs.

Accruals are not usually calculated in the accounts department for nominated suppliers, manufacturing suppliers, direct, nominated and labour only subcontractors. Instead they are normally calculated by the contractor's surveyor when completing the Cost Value comparison (see section 3.4.3.2 DFD No.3.4.3).

CALCULATE APPROVED SUBCONTRACT COSTS
Entry Type : Process

Specification :

This process deals with the following data : subcontractor liabilities, bills of quantities, subcontract orders. surveyor approved costs.

Between the external valuation and the receipt of the cost report from the accounts department, the surveyor should calculate the subcontract liabilities up to the agreed cost cut off date. These liabilities include for all known payments to which each subcontractor may be entitled and hence tend towards the 'worst case' costs. The liabilities when calculated are transferred to the 'Surveyors Approved Costs' sheet which forms part of the monthly 'Cost Value Comparison' report. Onto this sheet the surveyor is required to enter the names of each subcontractor (nominated, direct and labour only) and also the names of any manufacturing or nominated suppliers used on the contract to date. The net liability of each is then entered under the heading 'Surveyors Assessment of Liability', and the value of any discount is entered in the 'Discount' column.

CALL FOR BOQS
Entry Type : Data Element

CALL OFF REQUEST
Entry Type : Data Group

Composition:
CALL OFF REQUEST + CONFIRMATION OF CALL OFF

Comment: This 'call off request' is against a bulk order already placed.

CASH FLOW FORECASTS
Entry Type : Data Element

CASH FLOW MONITORING
Entry Type : Process

Specification:

This process deals with the following data: notification of value to be cert. & val. data, forecast payments sheet, certificate date, certified value, late payment chaser, contract turnover forecast.

The Surveying director should keep a forecast payments schedule which is used to monitor cash flow and the actual cheque value against the anticipated cheque value. The schedule will list all the current contracts within the region and against each lists the forecasted selling value of the contract for the appropriate month. These forecast selling values are abstracted from the 'Contract Turnover Forecasts' which the surveying director received at the beginning of the contract. The director also lists the valuation dates of each contract which he abstracts from the 'Schedule of Cost Dates'.

In addition to monitoring cash flow as above, the surveying director should monitor the actual selling value against the forecast selling value. As payment is received in any one month the actual payment value (selling value) is entered onto a 'Contract Turnover Forecast Sheet' against the 'Forecast Selling Value'. Any difference is then calculated and entered into the column headed 'Variance between Forecast/Actual'. Over time tends will emerge which will either prove the forecast to be correct or incorrect. The 'Turnover Forecast Sheet' should not be changed by the job surveyor would normally be required to provide explanation if the actual value did vary substantially for a prolonged period. The director would use the actual turnover figures in reviewing the expected regional turnover for a board meeting.

CERTIFICATE DATE
Entry Type : Data Element

CERTIFICATE DETAILS
Entry Type : Data Element/Local File

CERTIFIED VALUE
Entry Type : Data Element

CHECK DELIVERY
Entry Type : Process

Specification:

This process deals with the following data : delivery tickets, signed delivery tickets, in tray (delivery tickets) material orders, request to supply sheets.

Each delivery should be checked on site for content and

breakages. The delivery ticket should be checked for correctness and annotated appropriately if it cannot be fully verified. The quantity and quality should be physically checked and compared against the call-off note or the order. Any deficiencies in the delivery are noted on the delivery ticket, together with details of any legitimate waiting time experienced by the supply lorry.

Appropriate tests would be supervised by the Engineer.

CHECK MINUTES
Entry Type : Process

Specification:

This process will deal with the following data : minute corrections, official site meeting minutes, regional office meeting minutes file, site meeting minutes F., MD's minute files, site progress meeting file.

It is usual for the Architect to take and write up the meeting minutes. The Construction Manager and Agent check carefully that the official minutes record accurately and fairly all that was discussed during the meeting. The minutes are seen by the directors in the post and copies are subsequently filed both on site and in the regional office. The Construction Manager and Site Agent note any irregularities within the minutes and raise them at the next meeting for discussion under 'Matters arising from the minutes'.

CHECK PAYMENT AUTHORISATION
Entry Type : Process

Specification:

This process deals with the following data : normal payment authorisation, 'one off' payment authorisation, surveyors checks & calculations, 'one off' payment authorisation, normal payment authorisation.

Normally both the 'one off' and normal payment authorisation forms are checked and countersigned by either the Regional Surveying director or a senior Surveyor. In checking the payment details the director does not normally have the relevant application or surveyor's break-down and looks simply at the cumulative values. It is usual however, for the supporting documentation to be requested for payments which strike the director as being unusual and as such Surveyors are required to have prepared suitably details build ups or checks for possible analysis. Once checked and countersigned the authorisation forms are passed to the Accounts department.

CHECKED APPLICATION
Entry Type : Data Element

CHECKS
Entry Type : Data Group

Composition :
CHECKS + INFORMAL CONVERSATION + FORMAL CONVERSATIONS + MATERIALS ON SITE + SAFETY CHECK + QUALITY CHECK

CLIENT
Entry Type : Terminator

The term 'client' includes all public and private regular and irregular customers of the industry. That is the person requiring the project and paying for it.

CLIENT INFO
Entry Type : Data Group

Composition :
CLIENT INFO + LETTER OF INTENT/CONTRACT DOCUMENT + AGENDA + AGREEMENT ON FIRST MEETING

CO-ORDINATE SITE TEAM
Entry Type : Process

Specification:

This process deals with the following data : weekly programme, foremens' inputs, managers inputs, QS inputs, buyer's inputs, agreed course of action.

The co-ordinating of the site team is both a formal and an informal process. On the formal side there should be an internal review meeting held monthly which brings together the Site Agent, Construction Manager, Surveyor and job buyer (see Section 3.8 DFD No.3). The meeting should keep all parties aware of the position of the contract and includes a review of the contract in terms of its progress against programme and its financial status to date. The meeting should also look to the month ahead and examine the work to be undertaken and its associated resource requirements.

On a more informal basis there should be a constant liaison between the Agent and his Manager. The Manager should visit site at least once a week and in addition there will be telephone conversation usually daily to discuss the post and the immediate problems and requirements of the contract.

The Agent and General Foreman should work together in agreeing the daily workload which includes the drafting of the weekly programme each Friday. The weekly programme requires the agreement of the Foreman or gangers of the subcontractor's currently on site and is an important tool in the co-ordination of the workforce and the sequencing of the trades.

COLLATE QUOTATIONS
Entry Type : Process

Specification :

This process deals with the following data : trade files, quotations and quotations & comparison sheets.

As quotations are received back from subcontractors and suppliers they are passed straight to the tender estimator who edits them into trades as appropriate (for example, the separation of groundwork and excavation items).

The quotations are then passed to the comptometer operators who enter them onto analysis sheets to enable comparisons to be made. The level of detail shown on the analysis sheet depends upon the number of work items in each trade. Some trades are listed item by item, others page total by page total. As this stage there is a file for every trade which comprises the relevant tender enquiry abstract, the drawings and bill extracts despatched with the invitations to tender, the quotations received back and the analysis sheets.

COLLECT NECESSARY INFORMATION
Entry Type : Process

Specification :

This process will deal with the following data : scaffold schedule, plant requirement, plant enquiries, plant quotations, office visit report, site visit report, quotations, temporary works details.

The Planner is responsible for acquiring quotations for any scaffold, major plant and temporary works that may be required.

The scaffolding schedule, together with a letter of enquiry is dispatched as soon as the scaffold requirements have been defined. The major plant requirements are highlighted as the methods of construction are decided and the programmed activity durations calculated. Enquiries are despatched if time permits otherwise quotations can be obtained by telephone. The Plant Manager from a Company Plant Department may be consulted over the types of plant available to meet specific requirements but such enquiries are usually directed straight to the actual supply companies. Cranage is usually the principal plant requirement for which quotations are sought.

In conjunction with the tender estimator the planner usually makes a site visit and also a visit to the design consultant's office. The reports from these visits reflect the type of information which is gathered and the visits themselves enable a much fuller and clearer picture of the proposed contract to be built up.

COLLECT RELEVANT INFORMATION
Entry Type : Process

Specification :

This process deals with the following data : quotations, negotiation, invitations to tender, tender enquiry abstracts, subcontract register, supply register, completed tender enquiry abstracts, direct work abstracts, s/c names, tender documents, insurance premiums, insurance enquiry, site visit report, office visit report, supplier quotations and enquiry letters.

The enquiry clerk dispatches invitations to tender to subcontractors and letters of enquiry to material suppliers. The letter of enquiry is a standard letter which states the contract name and location and the date by which quotations should be returned. It requests the supplier to submit a competitive price together with cash settlement terms. Details of the materials concerned are either typed on the letter or attached in the form of an extract from the bill of quantities.

An invitation to tender is prefaced by a standard letter which is again typed and dispatched by the enquiry clerk. The content of the letter is set down by the estimator who enters the details relating to each invitation onto a standard letter. An insurance enquiry sheet is prepared by the chief estimator and set to the central group insurance manager situated at the company Head Office. The insurance manager subsequently advises the chief estimator of the premium required to effect adequate insurance on the contract.

The tender estimator and planner make a joint visit to

the site during the period between the dispatch and return of enquiries. The 'Site Visit Report Form' outlines the information collected during the visit although the report forms are not always completed. In addition they may also visit the consultants offices (architect or engineer) to examine detailed and working drawings not included in the tender documents. A consultants office report is completed during the visit and indicates the type of information that is being sought.

If only one set of tender documents has been received the estimator passes the documents to the planner once the enquiries have been dispatched (usually within four days of tender having been received).

COMPANY CASH FLOW STATISTICS
Entry Type : Data Element

COMPANY POLICY
Entry Type : Data Element/Local File

COMPARISON OF PRETENDER QUOTATIONS
Entry Type : Process

Specification :

This process deals with the following data : precise requirements, records of meetings, most competitive quotations, pretender quotes, comparison sheets, tender enquiry abstracts F.

Once the requirements are known the buyer re-examines the quotations received during the tender preparation. Normally the lowest tenderer will be given the opportunity to do the work after having satisfied the buyer and construction manager of his commitment and ability to produce work to a high quality and within agreed 'milestones'. It is likely that for the major trades the buyer would have been in contact with the most competitive subcontractor prior to tender finalisation to obtain any further margins that may be available. Such concessions are obtained on the understanding that should the company be awarded the main contract, the subcontractor would be given the work. It is therefore possible that a subcontractor may have already been chosen prior to the contract been awarded.

If after competition, no subcontractor is clearly the lowest, a detailed analysis of the quotations would be undertaken possibly with the assistance of a job surveyor and the most competitive quotations would be extracted.

COMPARISON SHEETS
Entry Type : Data Element/Local File

COMPETITIVE LATE QUOTATIONS
Entry Type : Data Element

COMPETITIVE QUOTATIONS
Entry Type : Data Element

COMPETITIVE QUOTATIONS & MEETING RECORDS
Entry Type : Data Element

COMPETITIVE SUBCONTRACTORS
Entry Type : Terminator

Comment :
This is a sub-set of 'sub-contractors' terminator.

COMPETITIVE SUPPLIERS
Entry Type : Terminator

Comment :
This is a sub-set of 'suppliers' terminator

COMPLETE BUYING SUMMARY SCHEDULE
Entry Type : Process

Specification :

This process deals with the following data : buying summary schedule (to 3.5), buying summary schedule F, bills of quantities spec., drawings, supplier order files, contract programme F.

Section 2.5.4 (in DFD No.2.5) described the role and procedures associated with completing the buying summary schedule. On contracts with short set up periods it is unlikely that the schedule was completed prior to work commencing on site. In such situations the agent is given a date by which the schedule should be completed and submitted to the buyer, this date is agreed during the contract commencement meeting.

Having submitted the schedule, the agent should receive a copy of it back with the dates by which each item should be requisitioned. These dates are entered onto the schedule by the buyer and account for the time taken to select a supplier and place the order and for the expected delivery time associated with the materials concerned.

COMPLETE COST VALUE COMPARISON
Entry Type : Process

Specification :

This process deals with the following data : adjusted true value, schedule of costs & adjustments, cost value comparison, previous CVC's file.

The Adjusted Costs are transferred onto the CVC against the six appropriate elemental heading of Labour, Materials, Plant, Direct Subcontractors, Nominated Subcontractors and Sundries. When totalled the six elemental costs reveal the net cost which when added to the 'overheads as tender' (a percentage addition) reveals the gross cost.

The adjusted true value column can only be completed fully if an analytical internal valuation has been prepared. As outlined in Section 3.4.1 in practice this is rarely done and thus the first figure to be entered in this column is the 'Total Net Value'. The total net value represents the internal valuation less the profit element contained within it. This is then added back in against the 'gross margin' heading to reveal the gross value. The gross cost is deducted from the gross value to reveal the net profit which is expressed as a value and a percentage. The 'target profit' is also calculated on the basis of the forecast profit margin established at contract commencement. This value is then listed below the net profit. As with the valuation analysis, the equivalent values from the previous month's valuation are also noted to reveal in the last two columns the months net profit.

COMPLETE CVC FORMS
Entry Type : Process

Full description see DFD No. 3.4.3.

COMPLETE CVC SUMMARY SHEET
Entry Type : Process

Specification :

This process deals with the following data : CVC summary files, monthly contract report/CVC files, monthly CVC summary, book cost & cash reconciliation, all current contract CVC reports.

Each month the surveying director will typically complete a 'Cost Value Comparison Summary' sheet. The information entered onto the summary is normally abstracted from the CVC's received in the previous month. All current contracts within the region and all contracts awaiting their final accounts to be settled should be listed on the summary sheet.

The surveying director should complete all sections across the sheet up to the columns headed 'Cash Reconciliation'. The accountant later completes the remaining sections and the two 'Balance Outstanding' boxes at the bottom of the page - See Figure 6.

The Adjusted Selling Value, Net Cost Price (NCP), overheads and profit for both the month and the contract to date are abstracted from the CVC entered under the appropriate headings. The profit is expressed as both a value and a percentage and is compared with the target profit percentage which is also listed. The current Architect's Certificate Value and the Retention are abstracted from the 'Certificate Details' section of CVC report sheet and listed next to the date of the corresponding valuation.

Having summarised the above mentioned details for all the contracts concerned the Selling Value, NCP, Overheads and Profit columns are totalled to the bottom of the sheet to reveal the regional profit for the month. The summary sheet is then passed to the Regional Accountant for the remaining sections to be completed.

The information entered onto the sheet is as follows:

1. 'Cash Received to Month End'
   The value of any cheques received during the month.

2. 'Since Month End or Forecasted'
   The value of cheques received after the month end but before the Accountant completes the sheet.

3. 'Balance'
   The value of cheques that have not been received. (The Surveying Director is required to give an explanation of any such values).

4. 'Certified Variance'
   The difference between the certified values and the adjusted selling values. There should be no 'over' variance as the QS cannot report the selling value as being over the certified value.

Cost Value Comparison Summary

Region ....

Month ....

Job No	Title	Period	Contract Value	This Month					To Date					
				Selling Value (Adj)	NCP	O/H as Tender	Profit (Net)	%	Selling Value (Adj)	NCP	O/H as Tender	Profit (Net)	%	Total %

W I P Make up is	
W I P Costs (as over)	
Add Cost provision (as over)	
= N C P as CVC	
Add Overhead and Profit (above)	
LESS Cash Received (above)	
= Balance Outstanding	

Fig. D6. Cost value comparison summary

Val Date	Certificate Value	Bal	Cash Reconciliation					Cost of W I P at month end	Cost Variance	
			Cash rec'd to month end	Since month end or	Bal	Cert Var			FAV	ADV
						Over	Under			

Make up of Balance outstanding	
Retentions (above)	
Add Cert unpaid (above)	
LESS Net Overheads *(Ie a Certified loss* Selling Value above	
= Balance Outstanding	

Fig. D6. Cost value comparison summary (continued)

5. 'Cost of Work in Progress at Month End'
   The 'book cost' (ledger cost) of each contract at the month end.

6. 'Cost Variance' - Favourable or Unfavourable.
   The difference between the book cost of work in progress at the month end and the surveyor's net cost price to the date of the valuation. The surveying director is required to give an explanation of any 'adverse' variances. These sometimes occur on final account contracts where costs may have been incurred since the last valuation.

The two boxes at the bottom of the sheet simply summarize the details on the upper part of the sheet to reveal the breakdowns of the 'Work in Progress' and the 'Balance Outstanding'.

COMPLETE POSTING SLIP
Entry Type : Process

Specification :

This process deals with the following data : arithmetically correct invoices, job number, cost centre number, date due, invoice and posting slip.

A posting slip is usually attached to an invoice and filled in. The appropriate contract and cost centre numbers are entered onto it, together with cost, discount and VAT. The "payment due date" is also entered onto the slip which is either the 'end of the month' or 'the date payment has to be made to obtain maximum discount'.

COMPLETE REMAINING SECTIONS OF REPORT
Entry Type : Process

Specification :

This process deals with the following data : CVC report (part), tender/valuation details, certificate details, forecast future losses/additions, cost value comparison report.

On the sheets of the CVC completed by the surveyor, there are various additional pieces of information to be entered.

At this stage some tender and valuation details are required. The contract sum, net cost, o/h in tender and profit values were established in the first weeks of the contract and should not change. The certificate number, valuation date and percentages of value expected and programme completed change from month to month and are self-explanatory.

Next a section entitled 'Certificate Details' is completed with the surveyor entering the pertinent details abstract from the latest Architect's Certificate. This is usually the certificate covering the current months valuation which should have been received by the time the CVC is reported. On this sheet the surveyor is required on a quarterly basis to complete the 'forecast of future losses/additions' section. Under this section the surveyor can inform the senior management of his assessment of any impending extraordinary losses/additions which are likely to occur on the contract. Typical items include any liquidated damages that may be payable if the contract is in delay, any serious making good which may need to take place toward the end of the contract, etc. i.e items for which staff at director level ought to be taking reserves.

COMPLETE SUBCONTRACTOR PROFIT/LOSS STATEMENT
Entry Type : Process

Specification :

This process deals with the following data : subcontract order details, bills of quantities, subcontract profit/loss statement, profit/loss statement F.

The buyer will usually enter the details of each subcontract order placed onto a subcontractor profit/loss statement. The statement, likes its materials counterpart (see section 3.5.1.2.6 in DFD No. 3.5.1.2) is normally passed to the company directors quarterly and is a means by which both the buying departments performance and the expected profit/loss on the contract in relation to subcontract works can be measured. Copies are also normally passed to the job surveyor and the construction manager quarterly simply as a statement of the current position and profit of the subcontract element of the contract.

As orders are placed the specialist trade subcontractors name and the order number are entered onto the sheet. The net order value and corresponding net bills of quantities values are then filled in and the resulting profit or loss calculated both as a value and as a percentage of the bq value. The cumulative profit/loss resulting from all orders placed to date is entered down the right hand margin of the sheet.

COMPLETE SUPPLIER PROFIT/LOSS STATEMENT
Entry Type : Process

Specification :

This process deals with the following data : order details, profit/loss statement F., bills of quantity details.

The buyer enters the details of each order placed onto some form of supplier profit/loss statement. This statement is passed to the directors quarterly and is a means by which the buying departments performance can be measured. It should also be passed to the job surveyor and construction manager quarterly as a statement of the position and profit of the materials supply content of the contract.

Each material should be itemised on the statement and the suppliers names and order number entered next to it. The statement compares the net order value for each material against the nett corresponding value on the bill of quantities. The difference is taken as the profit (loss) which is stated both as a value and a percentage. The cumulative profit (loss) of all the materials is recorded down the right hand margin of the form and is again stated as both a value and a percentage of the total order value.

Allowances are made for any buyers margins built into the tender at the tender adjudication meeting. Thus, the profits (losses) shown on the statement reflect actual buyers profits and not profits before deduction of expected margins.

COMPLETE VALUATION
Entry Type : Data Element

COMPLETE VALUATION ANALYSIS
Entry Type : Process

Specification :

This process deals with the following data : internal valuation, previous valuations, external valuation F., adjusted true value.

A valuation analysis is not always completed but when it is, requires the surveyor to list the 'QS Value', and the 'Adjusted True Value', for the elements of the valuation listed. The elements are broadly compatible with the format of the external valuation and the 'QS Value' column represents the externally agreed value whilst the 'Adjust True Value' represents the internal value derived by adjusting the external valuation in the manner outlined in Section 3.4.1 DFD No. 3.4.

The surveyor completes both the 'Total to Date' and 'Previous Total' columns (the latter being abstracted from the previous months CVC report). By deducting the latter from the former the surveyor then enters the difference in the 'This Month' columns.

COMPLETED ESTIMATE
Entry Type : Data Element

COMPLETED TENDER ENQUIRY ABSTRACTS
Entry Type : Data Element

COMPLETING PURCHASE PROCEDURE
Entry Type : Process

This process includes the following : photocopying and check invoices, complete posting slip, verify invoices, cost invoices, schedule pay dates, produce remittance advice and cheque, authorise and dispatch cheque.

COMPLETION OF PLANT-HIRE AGREEMENT
Entry Type : Process

Specification :

This process deals with the following data : payment and invoice rejection, invoices, external plant & transport return, internal plant invoice, external invoice verification, plant invoices, accounts plant & transport return F., invoice details, invoice verification, account copy order files.

The procedures used in checking, costing and paying invoices for plant are generally similar to those described in sections 3.5.1.4.1 to 3.5.1.4.7 in DFD No. 3.5.1.4., (the checking, costing and paying of materials invoices). The only differences are that invoices are verified against the external plant and transport returns and not the material returns and that the internal plant invoices are not verified by the invoice checkers as a copy is sent to site for agreement prior to being submitted to the accounts department.

COMPLEXITIES/ALTERNATIVE
Entry Type : Data Element

All possible methods of construction are considered and resources, gang sizes and plant capacities varied to appraise alternative solutions. The optimum solutions are sought for each activity and the planner may need to liaise with the managing director over the more complex decisions where viable alternatives exist.

CONFIRMATION OF 'CALL OFF' FILES
Entry Type : Data Element

CONFIRMATION OF CALL OFF
Entry Type : Data Element/Local File

CONFIRMATION OF INSTRUCTION
Entry Type : Data Element

CONFIRMATION OF INSTRUCTIONS
Entry Type : Data Element

CONFIRMATION OF INSTRUCTIONS FILES
Entry Type : Data Element/Local File

CONSIDER INFORMATION REQUIRED
Entry Type : Process

Specification :

This process deals with the following data : contract programme, information required, immediate information required.

Having commenced the preparation of the contract programme the construction manager considers the project information required both immediately and in the short term. Using a standard 'initial information required summary sheet' the manager requests the information required immediately. The summary is dispatched to the architect with a letter as soon as the contract is awarded. The letter emphasised the need for detailed information without which the planning function and site set up cannot be effective.

A further and more detailed examination of the information needed in the short term takes place with the manager and the site agent jointly reviewing the requirements. An 'information required schedule' is produced listing the bill of quantity reference numbers, descriptions of the items and the date required. In addition a schedule of PC sums and provisional sums is completed listing the dates by which nominations and details should be received together with the associated value of each item. Therefore two schedules are presented to the architect at the external precontract meeting (see 2.8 in DFD No.2).

CONSTRUCTION COMPANY OPERATION
Entry Type : Process

This is the system process in the 'context diagram'.

CONSTRUCTION MANAGER CVC F.
Entry Type : Data Element/Local File

CONSTRUCTION MANAGER'S F.
Entry Type : Data Element/Local File

CONSTRUCTION MANAGER'S FILE
Entry Type : Data Element

CONSTRUCTION MANAGERS F.
Entry Type : Data Element

CONSTRUCTION PROCEDURES
Entry Type : Process

This represents the construction phase, during which the contractor and the sub-contractors actually build the project.

CONSULTANTS REPORTS
Entry Type : Data Element

CONTRACT BILLS OF QUANTITIES
Entry Type : Data Element

CONTRACT COMMENCEMENT MONITOR
Entry Type : Data Element

CONTRACT COMMENCEMENT MONITOR F.
Entry Type : Data Element/Local File

CONTRACT COSTS
Entry Type : Data Element

CONTRACT COSTS
Entry Type : Data Element/Local File

CONTRACT DOCUMENT
Entry Type : Data Group

Composition :
CONTRACT PROGRAMME + AGREEMENT ON 1ST MEETING + IMMEDIATE INFO REQUIREMENT

CONTRACT DOCUMENT F.
Entry Type : Data Element/Local File

CONTRACT PROGRAMME F.
Entry Type : Data Element/Local File

CONTRACT DOCUMENTS - BILLS, SPEC, DRAWINGS
Entry Type : Data Element/Local File

CONTRACT FILES
Entry Type : Data Element

CONTRACT PRELIMINARIES
Entry Type : Data Element

CONTRACT PROCEDURES
Entry Type : Process

This process includes : agree weekly programmes, work supervision, work measurement, process interim applications, raise and dispatch payments.

CONTRACT PROGRAMME
Entry Type : Data Element

This will include any changes made during the 'secure contract' phase.

CONTRACT PROGRAMME F
Entry Type : Data Element/Local File

CONTRACT RELATED INFO
Entry Type : Data Element

CONTRACT TURNOVER FORECAST
Entry Type : Data Element/Local File

CONTRACTORS REPORT
Entry Type : Data Element

COPY ADDENDUM ORDERS BUYING
Entry Type : Data Element/Local File

COPY CONFIRMATION OF CALL-OFF INSTRUCTIONS
Entry Type : Data Element/Local File

COPY MATERIAL RETURNS
Entry Type : Data Element/Local File

COPY MRS FILE
Entry Type : Data Element

Comment : MRS = Material Returns Sheet

COPY ORDER FILE
Entry Type : Data Element/Local File

COPY ORDERS
Entry Type : Data Element/Local File

COPY ORDERS BUYING F.
Entry Type : Data Element/Local File

COPY SUBCONTRACT ORDER & APPENDICES
Entry Type : Data Group

Composition :
SITE COPY ORDER & APPENDICES + SUBCONTRACT PROFIT/LOSS STATEMENT

COPY TENDER ENQUIRY ABSTRACTS
Entry Type : Data Element

CORRESPOND & CERTIFICATE
Entry Type : Data Element

CORRESPOND RECEIVED ON SITE
Entry Type : Data Element

COST CENTRE NO.
Entry Type : Data Element

Comment : all costs will be charged to a cost centre number

COST INVOICES
Entry Type : Process

Specification :

This process deals with the following data : contract costs, verified invoice, invoice files, notice of payable invoice.

Once verified, invoices are normally batched in groups. Each batch will be totalled in terms of invoice value and VAT and the totals are entered into a control book. Information on the posting slip of each invoice within the batch is then entered onto a computer system which in turn gives a print out of the batch totals which can be compared with the control book entries to ensure postings are entered correctly. Unbalanced invoices will be rejected during re-checking.

A computer system should schedule invoices to be paid and allocate the costs to the appropriate contracts under the relevant cost code headings.

COST PRINT OUTS
Entry Type : Data Element/Local File

COST REPORT
Entry Type : Data Element

COST VALUE COMPARISON
Entry Type : Data Element

COST VALUE COMPARISON PROCEDURES
Entry Type : Process

Specification :

This process is concerned with the systems for establishing accurate costs, profit and general financial performance on a project. It deals with:

1. The system for the preparation of accurate internal valuations (external valuations less over measure etc).
2. The internal cost system for a project.
3. The transformation of data into the structure used on a CVC form.
4. The consultation process at the draft CVC report stage.
5. The CVC reporting system to directors.
6. The official CVC reporting information for the company accounting system.

COST VALUE COMPARISON REPORT
Entry Type : Data Element

COSTS
Entry Type : Data Element

CRITICAL DESIGN INFO
Entry Type : Data Element

CURRENT CVC REPORT
Entry Type : Data Element

CVC REPORT (PART)
Entry Type : Data Element

CVC REPORT TO SURVEYING DIRECTOR
Entry Type : Process

Specification :

This process deals with the following data : cost value comparison report, surveyors CVC, surveying directors CVC files, all current contract CVC reports, construction manager CVC F.

The surveying director receives the CVC reports for all current contracts. In presenting a CVC report to his director, the surveyor outlines the financial results and any significant variances from target. Only rarely would the surveyor be required to change the CVC figures as effectively they represent the true values. The surveying director would seek to establish the extent of any monies 'hidden' under the 'subcontractor's liabilities' section, i.e monies which may need to be paid out but which are unlikely to be claimed in full. If the construction manager has not yet completed his section of the CVC report (the operational report) the surveying director will remind him.

CVC SUMMARY FILES
Entry Type : Data Element/Local File

DAILY DIARY
Entry Type : Data Element/Local File

DAILY DAIRY FILES
Entry Type : Data Element/Local File

DAILY DIARY SHEETS
Entry Type : Data Element

DAILY DIARY SHEETS F
Entry Type : Data Element/Local File

DATE DUE
Entry Type : Data Element

DAYWORK RECORDS
Entry Type : Data Element

DAYWORKS
Entry Type : Data Group

Composition :
PRICE DAYWORKS + DAYWORKS

DAYWORKS FILES
Entry Type : Data Element/Local File

DECISION TO TENDER

Entry Type : Process

Specification :

This process deals with the following data : tender submission, accept invitation, invitation to tender, office visit report, tender summary sheet & tender documents.

Upon receipt of an invitation to tender a decision is taken by a director and the chief estimator whether to accept or decline. The decision is made after consideration of the following factors:

1. The current and future known workloads.

2. The type of work.

3. The location of the proposed contract.

4. Estimating Department workloads.

5. Expected competitors.

A letter accepting or declining the invitation is then sent.

DEFINED REQUIREMENT
Entry Type : Data Element

DELIVERY DETAILS
Entry Type : Data Element

DELIVERY TICKETS
Entry Type : Data Element

DESIGN INFO
Entry Type : Data Group

Composition :
DESIGN INFO + CRITICAL DESIGN INFO + CORRESPOND & CERTIFICATE

DESIGN INFO INSTRUCTIONS
Entry Type : Data Element

DESIGN TEAM
Entry Type : Terminator

The term design team includes architect, consultant engineers, chartered quantity surveyors and any 'project management' consultants.

DESIGN TEAM INSTRUCTIONS
Entry Type : Data Group

Composition :
DESIGN TEAM INSTRUCTIONS + ARCHITECT'S INSTRUCTIONS + CORRESPOND & CERTIFICATE

DETAILED/UPDATED INFO
Entry Type : Data Element

Advice on any changes (client, architect, other) is given to contract staff.

DETAILED ORDER PARTICULARS
Entry Type : Data Element/Local File

DETAILS OF SUBCONTRACTOR PERFORMANCE
Entry Type : Data Element

DETERMINE SCOPE OF SUBCONTRACT
Entry Type : Process

Specification :

This process deals with the following data : subcontract requisition, bills of quantities, specification, precise requirements, tender enquiry abstracts F, drawings, notice of s/c requirements.

Having received a requisition or having been made aware of the need to place a subcontract order a buyer proceeds to define the precise requirements in terms of the extent of the work, quality, etc. The tender enquiry abstracts produced at the pre-tender stage should list all the relevant bill pages and drawings that were available at the time of tendering. The buyer will have to establish whether the work content remains the same, and may need to check the more detailed working drawings now available.

DETERMINE SITE STRATEGY
Entry Type : Process

Specification :

This process deals with the following data : organising decisions, site layout plan, company policy.

Having been selected by the company directors the site team and associated construction managers and surveyors have to determine the site organisation in terms of office layouts. The individual responsibilities are normally laid down in the company policy, however, a measure of flexibility should exist to account for the differing personalities and skills of the staff.

The company should emphasize the need for teamwork and strong working relationships between personnel. It is most likely that surveyors and engineers are operationally responsible to the site agent or project manager. Surveyors however, usually remain functionally responsible to the surveying director and are based at the regional office. Exceptions to this often occur on large contracts when the site becomes near autonomous having site based surveyors, clerks and secretarial staff.

Site and office layouts are usually planned in detail especially when there is a restricted site. In such cases a detailed site layout plan would be completed to aid the optimum positioning of offices, etc. The layout of the offices is designed to aid communication between members of the site team which usually means they are situated in close proximity to each other.

DEVELOP A PROCUREMENT STRATEGY
Entry Type : Process

Specification :

This process deals with the following data : notification of urgent material requirement, material requisition, product data files, buying summary schedule, defined requirement, supplementary requirement to existing order, drawings, specifications, bills of quantities.

Upon the receipt of a material requisition a contract buyer will analyze the requirements to ensure they are fully and accurately specified. Any clarifications to the requisition that may be needed are sought directly from the site by telephone. Through reference to his copy of the contract bills of quantities (analytical) and specification; together with the central office copies of the relevant drawings; the buyer should be able to fully specify the requirements. A product data and component specification library that may be kept in a department further detail and describe materials required and enable a buyer to compare alternative produces available and alternative sources.

DEVELOP BUYING SUMMARY SCHEDULE
Entry Type : Process

Specification :

This process deals with the following data : tender information, activity dates, buying summary schedule.

One of the first tasks of the site agent is to complete the buying summary schedules. The schedule upon completion is passed to the buyer who uses it partly as a requisition and principally as a reference and control document to monitor the contract requirements throughout the duration of the works. The buyer also marks up on the schedule the dates by which he requires requisitions from site for those miscellaneous materials not covered on bulk orders. Site and the buying department keep a copy of the schedules throughout the contract and update it monthly at site meetings.

To prepare the schedule the site agent requires the bill of quantities and a copy of the programme which will not be a significantly altered before it is issued as the contract programme. A thorough analysis of the bills of quantities is necessary to enable the schedule to be completed effectively. This has the immediate benefit of quickly familiarizing the agents with the contract. The schedule covers all materials including nominated suppliers together with all provisional sums, direct subcontractors and labour only subcontractors.

DEVELOP CONTRACTS
Entry Type : Process

This process includes the following : arrange delivery, check delivery, record delivery, reconcile bulk materials.

DEVELOP TENDER INFORMATION
Entry Type : Process

All information received from the tender team is carefully re-examined.

DIRECT WORK ABSTRACTS
Entry Type : Data Element

DIRECTOR & MANAGER VISIT
Entry Type : Data Element

DIRECTOR F.
Entry Type : Data Element/Local File

DIRECTORS MONITORING OF MEETINGS
Entry Type : Process

Specification :

This process deals with the following data : contract files, managers minutes, buyers report.

Both the managing director and the surveying director receive copies of the managers and buyers minutes of the internal review meetings. The minutes provide a useful summary of the overall status of a contract and are retained by the directors as a ready source of information.

DISCUSS RESULTS W. MANAGER
Entry Type : Process

Specification :

This process deals with the following data : cost value comparison reports.

Having completed the CVC the surveyor meets with the construction manager for a short review of the month's results. The purpose of this meeting is to inform the manager of the financial position of the contract and to discuss generally the progress of the works. Should the CVC reveal poor results or adverse trends in profitability the meeting will seek to identify the problem and loss making sections of the work.

The CVC sheets, once completed should be signed by the construction manager, contract surveyor and one of the directors. The construction manager should add his own brief 'operational' report, which should report on the status of the following:

1. Programme
2. Relationships with external consultants
3. Quality of work
4. Claims position - on main contract
   - against subcontractors
5. Performance of company personnel/departments and subcontractors

6. General comments

DRAFT & DISPATCH LETTERS/REPLIES
Entry Type : Process

Specification :

This process deals with the following data : external files, office F., site F., director F., requiring replies.

Letters relating to production matters such as subcontractor performance problems are generally written by the manager who gets the necessary facts from the agent and surveyor. Letters to subcontractors of a financial nature and correspondence with the PQS are dealt with by the contract surveyor. Letters of a contractual nature would normally be written in conjunction with the directors who would decide the approach to be adopted. Office and site file copies are in general types for every letter and contractually significant letters would also be copied to the directors.

DRAFT & PLACE SUBCONTRACT ORDER
Entry Type : Process

This process includes : draft subcontract order, draft relevant appendix, obtain agreement to subcontract, set up subcontract payment procedures, complete subcontractor profit/loss statement.

DRAFT COST REPORT
Entry Type : Process

Specification :

This process deals with the following data : costs, accruals, cost report.

The accrued costs are entered onto a 'Final Reserve Entry Document' which gives a summarised breakdown of the accruals within each cost centre. These figures and the cost totals from the contract cost printouts are then transferred onto the contract cost 'Report Summary Sheet'. This sheet forms the face sheet of the monthly cost report and lists the months cost, accruals and total costs for each cost centre.

DRAFT PAYMENT AUTHORISATION
Entry Type : Process

Specification :

This process will deal with the following data : checked application, 'one off' payment authorisation, normal payment authorisation.

Having checked the applications and either agreed them or reduced them to an acceptable amount the contractors surveyor normally completes a payment authorisation form. Two types of form usually exist, a 'one-off' Payment Authorisation and a normal payment authorisation. Both forms are used for labour only, direct and nominated subcontractors and also manufacturing suppliers. The former is used for payments that are in some way different to the norm, for example payments that are including a VAT element or a 'contra charge'. The latter which is used in the majority of cases is normally for straightforward payments.

The content of the two forms is broadly similar, subcontractors name, job number, subcontract type (cost code number), cumulative gross certified value, pay on date and retention and discount percentages. The surveyor is usually required to calculate the net payment value in completing a one-off form, a task which is performed by the computer system for payments listed on the normal authorisation form. The 'pay on date' is used for the surveyor to convey to the accounts department the date the payment should be made. This may depend upon the terms of the relevant orders and the date of the covering architect's certificate. The authorisation forms are normally completed weekly, and provide the data necessary for the accounts department to be able to raise payments on the computer system.

DRAFT PRETENDER PROGRAMME
Entry Type : Process

Specification :

This process deals with the following data : tender documents, principal activities, methods/logic, programmed durations.

The pre-tender construction programme is a graphical representation of the proposed construction process. Drafted onto large programme sheets it is typically produced in bar chart form. (Networks may be drafted if it is a requirements of the tender). The activities on the method statement sheets are listed down the left hand columns of the programme sheet, together with any reports which may clarify the work content of the activity. The appropriate calendar dates are entered across the top of the programme and consist of the month, week commencement date and week number which is already printed onto the sheet. Activity durations are usually to the nearest whole week and the logic imposed by the construction methods determines the start and completion dates of each activity.

In addition to showing the activities the planner marks up the dates by which design information (drawings, bending schedules, etc) is required for each activity. The key to any symbols used in marking up the programme in this way is entered into the appropriate space at the top right of the sheet.

The level of detail given at pre-tender stage depends largely upon the nature of the work. Activities with durations of less than two weeks are usually left off the programme leaving between forty and sixty activities.

DRAFT PROGRAMME
Entry Type : Data Element

DRAFT RELEVANT APPENDIX
Entry Type : Process

Specification:

This process deals with the following data: detailed order particulars, subcontract order, terms of payment, s/c order & appendix, dayworks, programmed dates of commencement /completion.

Each subcontract order has one of the following appendices attached to it:

1 Appendix for direct subcontractors
2 Appendix for nominated subcontractors
3 Appendix for labour-only subcontractors

The buyer completes the appropriate appendix by reference to the subcontract requisition and the documentation and quotations forming the basis of the agreement. The appendices are used to schedule the details of the order. Reference is made to such matters as the expected start and completion dates, the value and terms of settlement of the order, dayworks, and the necessary information for calculation of any applicable price fluctuations. A supplementary appendix form is used to list any further terms and conditions not covered by the order and relevant appendix.

DRAFT SITE LAYOUT PLAN
Entry Type: Process

Specification:

This process deals with the following data: site visit notes, project information.

After visiting site the construction manager and agent may draft a site layout plan. The items usually marked on the plan are listed down the right hand margin of the site layout sheet and include services, offices, storage facilities, main plant, hard standings and bulk material stacking facilities. The plan is only produced for contracts with awkward or restricted sites when detailed planning of the layout is necessary.

DRAFT SITE ORGANIZATION STRUCTURE
Entry Type: Data Element

This element deals with the site management team structure & organisation.

DRAFT SUBCONTRACT ORDER
Entry Type: Process

Specification:

This process will deal with the following data: subcontract

quotations, summary of order details, site details, particulars of main contract, subcontract order.

Working from the subcontract requisitions sheet and details of the main contract, the buyer usually proceeds to complete an outline order form which is processed into the actual subcontract order.

On a typical order form the buyer enters the following summary details of the proposed order:

1 A general description of the works
2 The covering standard forms of subcontract
3 The bill of quantity and specification page numbers
4 The date of the subcontractors fixed/firm price quotation and its reference number
5 The total value of the quotation
6 Reference to any attached appendices

Particulars of the site and the main contract are normally given as detailed on the example order form.

The reverse side of a typical order lists the standing conditions of subcontract which unless specifically excluded elsewhere on the order form are applicable to each subcontractor.

The person who undertakes this task will vary from company to company. Normally it will be one of the following technical staff:

1 The buyer
2 The contractor's surveyor
3 The contractor's construction manager
4 The site agent/engineer

Although the person (and name) may vary from company to company, the process of placing the subcontract order varies very little.

DRAWINGS
Data Element/Local File

ENQUIRIES
Entry Type: Data Element

ENQUIRY & CONFIRM
Entry Type: Data Group

Composition:

PRICED DAYWORKS + INFORMATION REQUIRED SHEETS + CONFIRMATION OF INSTRUCTIONS + AGENDA + MINUTES + INFORMATION REQUIRED SCHEDULE

ENQUIRY LETTERS
Entry Type: Data Element

ESTABLISH ALL-IN RATES
Entry Type: Process

Specification:

This process deals with the following data: quotations & comparison sheets, direct work abstracts, subcontract quotations, subcontract comments, all in rates & quotations and tender documents.

The tender estimator prices little of the work himself. The only work sections usually priced when time permits are groundwork, carpentry and brickwork. To build up unit rates for these items the estimator has first to establish all-in-rates for labour, plant and materials.

For labour, an all-in hourly rate is calculated and is the sum of the operatives/tradesmen basic wages, associated emoluments, statutory costs and variable costs.

Major items of plant are charged under the project overheads (preliminaries). For small plant, however, (e.g. mixers), the estimator calculates all-in rates which include the basic hire charge, running costs, erection and transport costs. Hire rates are obtained directly from the company plant manager.

Material all-in rates are based on the quotations received back from suppliers which in the case of the more complex bulk orders will have been extended onto analysis sheets. Before quotations can be meaningfully compared the estimator translates them into all-in rates by making allowances for the associated costs of delivery, price fluctuations and any onerous supply conditions forming part of the quotation. This may involve consultations with the chief estimator and buyer whose experience and knowledge of the market place may be of great benefit. Having calculated all-in rates the estimator selects the keenest reliable supply quote for inclusion in the tender. Late quotations are monitored to check that the chosen quotation remains the keenest.

Eight to ten days before the tender has to be submitted the estimator selects a subcontractor for each trade. (usually the one who has submitted the most competitive tender). This requires a close inspection of the analysis sheets which have been prepared by the comptometer operators. To enable a meaningful comparison, the tenders have to be adjusted by means of allowances for risk, attendances and price fluctuations. This is normally done in liaison with the buyer especially on the larger subcontracts.

ESTABLISHED UNIT RATES
Entry Type: Data Element

ESTIMATED RATES OF BASIC PAY
Entry Type: Data Element/Local File

ESTIMATING
Entry Type: Data Element/Local File

ESTIMATING PROCEDURES
Entry Type: Process

This process is to determine the likely cost of all proposed work, including building up unit rates, prior to the preparation of the tender bid.

EXAMINE DOCUMENTS
Entry Type: Process

Specification:

This process deals with the following data: tender files, tender documents, tender summary sheet, trade files, copy tender enquiry abstracts and tender enquiry abstracts.

The estimator selected to prepare the tender conducts a detailed examination of the tender documents to gain an overall understanding of the project. A page by page analysis of the bills of quantities reveals the items for which quotations are required and these items are transferred onto tender enquiry abstract sheets. One abstract sheet is completed for each trade or material and the BQ and specification page numbers relating to each enquiry are listed. Also listed are the reference numbers of any drawings which he considers should be dispatched with the invitations to tender. Having completed the abstracting of information the estimator runs through each sheet and enters the name and address of reliable subcontractors and suppliers from whom he is likely to get quotations back. The estimating department has its own directory of subcontractors and suppliers but regularly sends invitation out to the same companies with whom a good relationship has been built up in the past. The completed tender enquiry abstracts are passed to an enquiry clerk who photocopies the necessary bill pages and prepares the invitations to tender.

EXTERNAL CORRESPONDENCE
Entry Type: Data Element/Local File

EXTERNAL FILES
Entry Type: Data Element/Local File

EXTERNAL INVOICE VERIFICATION
Entry Type: Data Element

EXTERNAL PLANT & TRANSPORT RETURN
Entry Type: Data Element

EXTERNAL PRECONTRACT MEETING
Entry Type: Process

Specification:

This process deals with the following data: client info, contract programme, agenda, information required, detailed /updated info.

The company personnel who would attend this meeting are:

1 Managing Director
2 Surveying Director (in MD's absence or on large contracts)
3 Estimator
4 Construction Manager
5 Site Agent
6 Quantity Surveyor

A company will have its own standard pre-contract meeting agenda, which in the absence of an agenda from the architect, would be used at the meeting. In cases where the architect does present an agenda the contractor's team would use their own as a check list to ensure all items are discussed. The contract programme and information required schedules would be taken to the meeting if they are completed. If not, they would be presented at the first contract progress meeting.

It is usual for the company to receive copies of the bills of quantities, specification, annotations, etc, and detailed drawings for the early work sections, together with related reinforcement schedules and any further information necessary to enable urgent orders to be placed.

It is the aim of the meeting to introduce all members of the contractor's team to the consultants representing the client, and to establish the procedures of communication and administration between the parties for the duration of the contract.

EXTERNAL PROGRESS MEETING
Entry Type: Process

This process deals with the systems associated with the provision of information for the client's advisers concerning the status of the project.

EXTERNAL VALUATION
Entry Type: Data Element

This data element represents an accurate assessment of the work undertaken during the last period, and will be used in the 'cost value comparison process' to determine:

1 performance against budget
2 performance against predicted project turnover
3 performance against profit forecast

EXTERNAL VALUATION F.
Entry Type: Data Element/Local File

EXTERNAL VALUATION PREPARATION
Entry Type: Process

This is the system for the preparation of the external valuation.

EXTERNAL VALUATION PROCEDURES
Entry Type: Process

Prior to meeting the PQS the contractor's surveyor will prepare a draft valuation. At an agreed date each month the two quantity surveyors will meet and analyze the contractor's draft valuation. This will form the basis of the recommendation by the PQS to the architect for the monthly issuing of the architect's certificate.

This process covers all activities associated with getting paid for work done, and will include the co-ordination of payment applications from all sub-contractors and suppliers.

EXTRA ORDERS
Entry Type: Data Element

EXTRACT INFORMATION
Entry Type: Process

Specification:

This process will deal with the following data: tender summary sheet, tender documents, bulk quantities, scaffold schedule plant requirement.

The tender planner receives a copy of the completed 'Tender Summary Sheet' and a set of tender documents. (If only one set of tender documents is received by the company, the planner should have access to that set during the period between the dispatch of enquiries and the return of quotations).

The planner runs through the bill of quantities and extracts all the bulk quantities. In addition, as his understanding of the project increases the plant and scaffold requirements will become evident. The scaffold requirements are abstracted onto the scaffold schedule and the major plant requirements listed to enable enquiries to be subsequently dispatched.

Any temporary works requirements may be referred to the

engineering department for the design of especially difficult propping and support systems.

The preambles and specification are read to check, e.g. for restrictions in concreting bay sizes, the type of concrete mixes to be used, and any other production related constraints.

FINAL ACCOUNT
Entry Type: Data Group

Composition:

FINAL ACCOUNT + SIGNED FINAL ACCOUNT

FINAL ACCOUNT AGREEMENT FORM
Entry Type: Data Element

FINAL ACCOUNT NEGOTIATIONS
Entry Type: Data Element

FINAL PAYMENT
Entry Type: Data Element

FINAL SELECTION
Entry Type: Process

Specification:

This process deals with the following data: competitive quotations & meeting records, subcontract quotation.

The selection of a subcontractor is likely to be straightforward after the detailed analysis of the quotations, negotiations and the subsequent meeting. The buyer, construction manager and surveyor ideally come to a joint decision although in cases where there is disagreement the construction manager would normally make the final decision.

Upon selection of the subcontractor a second subcontract requisition sheet is normally completed. The buyer or surveyor would enter the financial details, order value, bill of quantity value, and the resulting expecting profit (loss), together with a list of the documentation to be included in the order (quotations, letters, etc). The requisition sheet is then passed to the construction manager to enter the relevant programme details which will also form part of the order. The commencement and completion dates and any sectional completion dates are stated, together with the overall period for the subcontract works and the anticipated start date. The construction manager is typically required to sign the sheet as authorization for the order to be placed.

FINAL VALUE TO BE CERTIFIED
Entry Type: Data Element

This represents the notification to the contractor of the sum being recommended to the architect by the PQS for an architect's certificate.

FINALIZATION
Entry Type: Process

Specification:

This process deals with the following data: work record & measurement, final account, signed final account, final account negotiations, final account agreement form, final payment, main contract final account files, subcontractor files.

Upon completion of the subcontract works the subcontractor is normally requested to submit a final account. Having received this the surveyor will analyze it and checks it against his own measurements. It is frequently necessary to alter sections of the account which cannot be substantiated, notably the value of works in connection with variation orders. Negotiations may be held with the subcontractor over disputed items in the account and when necessary meetings are arranged with senior staff of the subcontractor to produce a fair settlement as quickly as possible. Protracted correspondence over such matters should be discouraged as it is time consuming and rarely results in agreement.

Once in a position of agreement, the contractor's surveyor sends the subcontractor a statement of final account form. The form states the final details of the account and requests a signature of agreement from the subcontractor. Upon its return the surveyor will release any final payments due to the subcontractor in the same manner as described in section 3.5.2.4.4 in DFD No. 3.5.2.4 (payment of interim applications). Retention monies would still be retained at this stage and released at the due times set down in the main contract or if different the subcontract.

FLUCTUATION TOTALS
Entry Type: Data Element

FORECAST CONTRACT CASH FLOW
Entry Type: Process

Specification:

This process deals with the following data: contract bills of quantities, contract programme F., agreed weekly programmes, company cash flow statistics, contract turnover forecast, cash flow forecasts.

During the first week of a project the contract surveyor is

generally required to analyze the bills of quantities and ascertain where the overhead and profit is allocated. Upon the finalization of the master contract programme he is also required to prepare a revised preliminary cost control document which should reflect any changes in the preliminaries content of the tender resulting from the revision or updating of the pre-tender programme. That is, he should prepare a statement of how the preliminaries will now be expended as a result of any changes by the construction manager and site agent in the methods of construction to be adopted.

Having prepared the above the Surveyor is then normally expected to produce a cash flow forecast for the contract based upon the most appropriate of the following alternative methods:

1 pricing the contract programme

2 applying company statistical data about a standard cost distribution pattern to the contract sum.

Generally the approach to be adopted will be decided by the surveyor in liaison with his surveying director and will depend largely upon the nature and size of the contract work.

FORECAST FUTURE LOSSES/ADDITIONS
Entry Type: Data Element/Local File

FORECAST PAYMENTS SHEET
Entry Type: Data Element/Local File

FORECASTING RESOURCE REQUIREMENT
Entry Type: Process

Specification:

This process deals with the following data: buying summary schedule F., contract programme F., plant schedule F., site diaries, managers' prompts, surveyor's prompts, info requirement, resource requirement, information required sheets, information requirement, plant requirement, S/C requirement, materials requirement, resource forecast.

The role of a site team in forecasting resource requirement is seen to be the early consideration of all factors that may lead to delay on the contract. The need to place orders in good time and to raise requisitions for materials and plant long in advance of them being required on site is obvious but it can easily be overlooked in the heat of day to day problem solving, coordinating, planning, etc. The site agent is principally responsible for ensuring that resources and design information are available when required although the construction managers and surveyors who visit site regularly are able to take a more subjective view and hence have a major role to play in this aspect of site management. The principal matters necessitating constant forethought include:

1. Resource Requirements

The short and long term requirements for general labour and directly employed operatives and tradesmen.

The need to requisition materials.

The need to call off onto site bulk order materials.

The plant requirements short term and whether requisitions ought to be raised.

The need for subcontract orders to be finalized.

2 Information Requirements

The drawings and schedules needed from the design team and the dates by which they are required.

The need for nominated subcontractors to be selected by the architect to enable orders to be placed.

Detailed planning and programming provides an agent with his resource requirements. Labour and plant requirements can be gleaned from planners' original method statements and from experience. Both these resources have to be considered when drafting the weekly programmes. The daily requirements for direct labour and plant should be entered onto the bottom of the sheet. The 'buying summary schedule' must be constantly at hand to prompt the requisitioning of materials. The schedule should be fully reviewed and updated during the monthly site review meetings (see section 3.8 in DFD No. 3) during which a following month's requirements will be analyzed. The marking up of the master programme on a weekly basis stimulates consideration of the need to get subcontract orders placed or to call subcontractors onto site to commence work. It also prompts the agent to consider the information requirements which can be marked upon the programme with arrows pinpointing the dates that detailed design information is required. Urgently required information would be requested on the 'Company Information Required Sheet' or would be recorded onto the 'Information Required Schedule' to be raised at the next external progress meeting (see section 3.7 in DFD No. 3)

It is useful for the 'Daily Diary Sheets' to have a section at the bottom that is required to be completed on the last day of the week only. This includes a reference to the 'Buying Summary Schedule' which is central to the company's monitoring and control of future material and subcontract requirements. The question requires the agent or foreman to state whether the schedule has been reviewed during the week.

FOREMAN'S INPUTS
Entry Type: Data Element/Local File

FOREMAN'S INPUT TO PROGRAMMES
Entry Type: Data Group

Composition:

FOREMAN'S INPUTS TO PROGRAMMES + ACCEPTANCE OF ORDER

FOREMAN/AGENT'S RECORDS
Entry Type: Data Element/Local File

FOREMEN'S FILES
Entry Type: Data Element/Local File

A general recording system for important daily information

FOREMEN'S INPUTS TO PROGRAMMES
Entry Type: Data Element

FOREMEN'S INPUT
Entry Type: Data Element

Input from foremen employed by specialist trade contractors.

FOREMEN'S INPUTS
Entry Type: Data Element

FORMAL CONVERSATIONS
Entry Type: Data Element

This data element covers the transmission (collection) of more formal data, so as to ensure the effectiveness of the site management system. This would include data such as 'verbal site instructions' by an architect and contractual information to or from specialist trade contractors.

HANDOVER
Entry Type: Process

Specification:

This process deals with the following data: tender information, staff appointments, subcontract quotations & comparison sheet, contract commencement monitor.

Upon receipt of a letter of intent from the client or notice of the award of the contract a job handover meeting is arranged. The meeting usually takes place within four days of having secured the contract. In the days before the meeting the directors select the site team (see 2.3 in DFD No. 2) and the tender estimator prepares three copies of the tender file. (This contains all pre-tender information). The construction manager and the Agent begin to develop the pre-tender programme which they obtain from the planner.

The objective of the handover meeting is for the precontract service department to hand over all information on a new contract to the Production Department so that work can proceed.

Present at the meeting are:

i Managing Director - Chairman
ii Surveying Director*
iii Tender Estimator
iv Tender Planner
v Construction Manger
vi Site Agent
vii Quantity Surveyors
viii Job Buyer

* On major contracts and in the Managing Director's absence

HEAD OFFICE
Entry Type: Terminator

This terminator composes the following sub-terminators: regional office, safety officers and in-house plant dept.

HEAD OFFICE ENQUIRY
Entry Type: Data Group

Composition:

HEAD OFFICE ENQUIRY + INSURANCE ENQUIRY

They are enquiring about the legal & insurance aspects of the tender.

HEAD OFFICE INFO
Entry Type: Data Group

Composition:

SAFETY REPORTS + PROFESSIONALS' INPUT + MANAGERS' INPUT + PLANT INVOICE + VERIFICATION

HIRE LABOUR
Entry Type: Process

Specification:

This process deals with the following data: project information, preliminaries schedule, contract programme F., estimated rates of basic pay.

Labour, other than labour only subcontractors, is arranged

and brought onto site by the site agent in conjunction with his construction manager. A company would typically carry a small number of its own operatives. If at any time they are to be moved to a new site they will be issued with a 'notice of transfer sheet'.

Labour agencies may be approached especially when the volume of work for labourers will be erratic (this avoids a hire and fire situation). Alternatively the agent will advertise for labourers and hire on the basis of a successful interview.

HIRE PLANT
Entry Type: Process

Specification:

This process deals with the following data: requisitions, contract preliminaries, plant order, agreed hire rates, project information.

In addition to receiving plant requisitions the plant manager may attend the internal contract commencement meeting during which the plant requirements for the contract would be discussed (see section 2.7 in DFD No. 2). The plant manager is responsible for overseeing all plant and equipment requirements other than the tower crane which is hired from site by the construction manger and the site agent.

The plant department carries a stock of mechanical plant which is largely small and medium sized equipment. The plant manager decides whether to supply company plant or externally hired plant and arranges the order as appropriate. He also keeps a plant movement board to monitor the location of plant throughout the region. The plant manager decides whether to hire or purchase plant when requisitions arrive. The decision depends upon the current stock of that item, the approximate duration of the plant requirement and expected future demand. A plant order form is drafted and dispatched by the plant manager irrespective of whether the plant is to be hired or purchased.

HIRE PLANT ACCEPT PROCEDURE
Entry Type: Process

Specification:

This process deals with the following data: in tray, site copy orders, plant schedule F., requisition/ termination F., prompt to off hire, terminate hire notes, external plant & transport return, notice of off hire, plant delivery ticket.

Plant delivered to site is normally accompanied by a delivery ticket. The plant is usually checked to ensure it is in working order and any superficial defects would normally be noted on the delivery ticket. Upon satisfaction that the plant is as requested and in working order the agent or foreman will sign the delivery ticket retaining a site copy for entry onto the next external plant and transport return.

Any breakdowns are notified to the plant manger who arranges with the plant hire company for a maintenance visit or a replacement item of plant. The plant manager visits each site approximately once every two months. The visits usually take the form of a walk round site with the agent during which the manager is assessing how well the equipment is being looked after and ascertaining future requirements. A measure of the reliability of plant is also obtained from the visits.

Any externally hired plant is generally recorded weekly onto an external plant and transport return. The return is completed by the agent or foreman and submitted to the accounts department weekly. The form details the plant, its hours on site and related order numbers together with its related delivery or return ticket number. Note is typically made by the agent of any hours during which the machine was broken down.

Internally hired plant and equipment such as site hutting is normally invoiced directly by the plant manager to the accounts department. A copy of the internal invoice is dispatched to the site by the plant manager for checking by the agent. Any mistakes or wrong allocations are notified back to the plant manager as soon as possible by the agent for rectification of the invoice.

HIRE RATES
Entry Type: Data Element

HIRE RATES + QUOTATIONS
Entry Type: Data Element

IMMEDIATE INFO REQUIREMENTS
Entry Type: Data Element

IMMEDIATE INFORMATION REQ.
Entry Type: Data Element

IMPLEMENT 'ACTION' NOTES
Entry Type: Process

Specification:

This process deals with the following data: minutes & action notes, site diary, site progress meeting file.

Should action be needed on matters arising from the progress meeting the construction manager and site agent decide who will be responsible and delegate respectively. Matters not requiring immediate attention are recorded in the site and managers' diaries for future prompting.

It may be necessary to fill in confirmation of verbal instruction forms to confirm any instructions given during the meeting. Following the normal procedure (see Section 3.6.4 in DFD No. 3.6) the site agent will be responsible for completing them.

IMPLEMENTATION OF ACTION ON LESSONS LEARNED
Entry Type: Process

Specification:

This process deals with the following data: action notes, new/revised instructions, job review meeting files, reminder to requisition of materials, reminder to subcontractors, reminder to requisition of plant, minutes agenda & action notes.

The construction manager normally monitors informally the implementation of action notes resulting from the meeting. Typical action to be taken as a result of the meeting includes the requisitioning of materials, subcontractors and plant together with the off hire of unnecessary plant. It may also be necessary for the agent to change slightly the methods of construction or the sequencing of trades and he may have to increase or reduce the number of labourers on site. The buyer may need to hurry through certain orders and gather any information requested by the agent or manager. The surveyor may be requested to analyze in greater depth the worrying or adverse trends in profitability and to pinpoint those areas losing money.

IMPLEMENTED SAFETY REPORT
Entry Type: Data Element

IN TRAY
Entry Type: Data Element/Local File

IN TRAY (DELIVERY TICKETS)
Entry Type: Data Element / Local File

IN-HOUSE PLANT DEPT
Entry Type: Terminator

This is a sub-department of 'Head Office' terminator

IN-HOUSE PLANT DEPT NOTICES
Entry Type: Data Element/Local File

INFO FILES
Entry Type: Data Element

INFO FOR HEAD OFFICE

Entry Type: Data Group

Composition:

MONTHLY CVC SUMMARY + SAFETY NOTES + INTERNAL MEMOS + TELEPHONE CALL + PLANT INVOICE

Most of this information for head office consists of reporting for control activity by senior managers, for example cost control by means of the 'cost value comparison' (CVC) report.

INFO REQUIREMENT
Entry Type: Data Element

INFO TO S/C
Entry Type: Data Group

Composition:

NEGOTIATIONS /MEETINGS + FINAL ACCOUNT NEGOTIATIONS + WEEKLY PROGRAMMES /AGREED ACTION + INSTRUCTIONS TO S/C + PAYMENTS + FINAL ACCOUNT AGREEMENT

INFORMAL CONVERSATIONS
Entry Type: Data Element

This data element covers the transmission (collection) of 'live' and 'topical' data necessary for the site management systems to operate effectively.

INFORMATION REQUIRED
Entry Type: Data Element

The client is advised of all information requirements (with dates) if the tender programme is to be adhered to.

INFORMATION REQUIRED FILE
Entry Type: Data Element/Local File

INFORMATION REQUIRED SCHEDULE
Entry Type: Data Element

INFORMATION REQUIRED SHEETS
Entry Type: Data Element

INFORMATION REQUIREMENT
Entry Type: Data Element

INFORMATION REQUIREMENTS
Entry Type: Data Element

INITIAL ACTIVITIES DATES
Entry Type: Data Element

INITIATE CASH FLOW MONITOR
Entry Type: Process

Specification:

This process deals with the following data: P.Q.S files, architect P.Q.S & client files, valuation dates, schedule of cost dates, Q.S director files.

At the beginning of the contract the surveyor is normally required to agree with the pqs the valuation dates for as long into the contract as possible. Prior to doing this he should liaise with the surveying director to agree the period of the month that would be preferable to the company for the valuations to take place. The surveying director normally keep a contract valuation schedule which lists the valuation dates of all the current contracts within the region. In consulting this schedule the surveying director recommends a suitable time of the month bearing in mind his desire to spread both the surveyor's workload and the cash flow into the company across the month. Having established the preferred dates the actual valuation dates are fixed with the pqs as near as possible to them.

Having agreed the valuation dates, it is sensible for the surveyor to complete a schedule of cost dates which should itemize the following for each month:

i the valuation date

ii the week ending date to which the month's costs should be calculated

iii the date the surveyor requires the month's cost report from the accounts department.

This schedule is usually forwarded to the accounts department to give notice of the dates by which they are required to produce monthly cost reports for the contract. In addition, it is wise if a copy is sent to the surveying director to inform him of the agreed valuation dates. These dates are then transferred onto his contract valuation schedule and provide a reference from which the director can subsequently monitor the receipt of related architect's certificates and payments.

INITIATE PROCUREMENT
Entry Type: Process

Specification:

This process deals with the following data: contract programme, variations/ instructions F., review meeting notes, buying summary schedule, material requirement, site requisition file, construction managers F., material requisition.

Following the award of a contract to a company the buyer will have placed orders for the majority of bulk materials and for the miscellaneous one-off materials required on site in the initial months. The procedures adopted during that period are outlined briefly in sections 2.5.5 and 2.6.1 in DFD No. 2.5 & 2.6. Many materials remain unordered, and the procedures outlined below detail the methods used by a typical company for requisitioning, procuring and administering them.

During the construction phase of a contract the buyer would not normally commence the procurement of materials without receiving an appropriate material requisition. The exception may be when an urgent requirement becomes apparent too late and is either telephoned through from site or noted by the buyer during the monthly internal review meeting (see section 3.8 in DFD No. 3).

Material requisitions are normally completed by the site agent or on a large contract by the site clerk. There are several controls which serve to prompt the agent into requisition materials in good time. The principal document is generally the buying summary schedule which, having been completed at the start of the contract (see 2.5.4 in DFD No. 2.5), itemizes all the material requirements on the contract and the date by which requisitions should be raised to ensure orders are placed in sufficient time to avoid delays. The buying summary schedule should be consulted at least weekly by the agent who would normally indicate on the last daily diary sheet of each week whether he has or has not reviewed it during that week. As discussed in section 3.8 on DFD No.3 there is usually a monthly internal review meeting held on site to which the buyer is in attendance. During this meeting the material requirements for the following month can be analyzed and action such as the 'requisitions to be raised' agreed. The agent should receive minutes from the buyer indicating the matters discussed and agreed during the meeting and itemizing any action to be taken in the coming month. In addition the contract programme should serve to remind the agent of the work activities to be undertaken in the short term and prompts the requisitioning of associated materials.

INITIATING PROJECT CONTROLS
Entry Type: Process

This process has been further divided into some sub-processes.

INSTRUCTIONS
Entry Type: Data Group

Composition:

WEEKLY PROGRAMME + INSTRUCTIONS + INFO TO S/C + ARCHITECT CERTIFICATE + VARIATION ORDER + DESIGN TEAM INSTRUCTIONS + DESIGN INFO + VERBAL INSTRUCTIONS + PROJECT INFORMATION + CONSULTANTS REPORTS + OFFICIAL SITE MEETING MINUTES

This includes instructions from the client and the design consultants via the architect to the main contractor, but also information from the main contractor to sub-contractors.

INSTRUCTIONS TO S/C
Entry Type: Data Group

Composition:

ACCEPTANCE CHASING LETTER + SUBCONTRACT ORDER & APPENDICES + VALUATION DATES + AGREED ACTION

INSTRUCTIONS TO SUPPLIER
Entry Type: Data Element

Usually in accordance with the contract programme

INSTRUCTIONS TO SUPPLIER FILES
Entry Type: Data Element/Local File

INSURANCE ENQUIRY
Entry Type: Data Element

INSURANCE PREMIUM
Entry Type: Data Element

INTERIM APPLICATIONS
Entry Type: Data Group

Composition:

NSC'S INVOICES + NS'S INVOICES

Comments:

NSC'S = Nominated Sub-contractors
NS'S = Nominated Suppliers

INTERIM INVOICE F.
Entry Type: Data Element/Local File

INTERNAL ASSESSMENT OF S/C PROGRESS
Entry Type: Data Element

INTERNAL CONTRACT COMMENCEMENT MEETING
Entry Type: Process

Specification:

This process deals with the following data: contract commencement monitor, project information.

It is normal that directors have an internal contract commencement meeting for every contract. However, small or less complex contracts may not have such a meeting.

Given sufficient time to set the contract up (the company should request four weeks, although will seldom get that long) the meeting should take place one week after the contract handover meeting and ideally prior to the external precontract meeting. The contract commencement monitor acts as the agenda with the purpose of the meeting being to review the checklist of activities and responsibilities listed within it.

Present at the meeting would be:

1 Managing Director

2 Surveying Director (on major contracts and in MD's absence)

3 Site team
- Planner
- Regional Engineer
- Security Officer
- Plant Manager
- Buyer

The meeting is chaired by the managing director or in his absence the surveying director and will typically last half a day. It is the first full meeting of the site team. Check lists are run through and the roles and responsibilities of the different members of the site team agreed. The roles and responsibilities of each member of the team are outlined and the contract discussed from the aspect of each activity. Specialist contributions into the meeting from the regional engineer, security officer, safety officer, etc, will be requested by the directors when aspects of the contract raise specific problems in any particular area of their responsibilities.

The meeting serves to co-ordinate the large number of activities which take place during the set up of a contract and to ensure that none are overlooked. It also ensures that each member of the site team is conversant with the contract and understands their own responsibilities together with those of their colleagues. Should there not be time to hold the meeting before work starts on site, it will be held at the first opportunity.

INTERNAL MEMO FILES
Entry Type: Data Element/Local File

INTERNAL MEMOS
Entry Type: Data Group

Composition:

INTERNAL MEMOS + DAILY DIARY SHEETS + PETTY CASH VOUCHER

INTERNAL PLANT INVOICE
Entry Type: Data Element

INTERNAL REVIEW MEETING MINUTES
Entry Type: Data Element/Local File

INTERNAL SITE REVIEW MEETING
Entry Type: Process

This process covers the systems associated with the setting of project budgets and performance targets, their review and revision. Also the creation & implementation of action plans to change or improve performance.

INTERNAL VALUATION
Entry Type: Data Element

INVITATION TO TENDER
Entry Type: Data Element

The client with professional adviser will invite a number of contractors to attend a 'pre-qualification' interview. From those interviewed the client plus his team, will select a small number of companies who will be asked to produce a detailed tender.

INVITATIONS TO TENDER
Entry Type: Data Element

The main contractor will invite a number of domestic sub-contractors with a suitable performance track-record, to tender for work on the project. They will usually be asked to tender on the basis of the terms and condition of the main contractor's form of sub-contract agreement.

INVOICE & POSTING SLIP
Entry Type: Data Element

INVOICE DETAILS
Entry Type: Data Group

Composition:

INVOICE QUERY + INVOICE DETAILS

INVOICE FILES
Entry Type: Data Element

INVOICE FILES
Entry Type: Data Element

INVOICE PAYMENT/ REJECTION
Entry Type: Data Element

INVOICE QUERY
Entry Type: Data Group

Composition:

INVOICE QUERY + INVOICES + DELIVERY TICKETS + WORKS/YARD VISIT NOTES + MATERIALS IN WORKS + MATERIALS ON SITE + NEGOTIATIONS

INVOICE VERIFICATION
Entry Type: Data Element

INVOICES
Entry Type: Data Element

This concerns invoices from suppliers

INVOICES RECEIVED BUT NOT COSTED
Entry Type: Data Element / Local File

ITEM COST FILES
Entry Type: Data Element/Local File

JOB NUMBER
Entry Type: Data Element / Local File

JOB REVIEW MEETING
Entry Type: Process

Specification:

This process deals with the following data: buying summary, agenda, current CVC report, s/c lab, plant & mat reports, programme v progress report, action notes, minutes & action notes.

This meeting is usually held monthly on site and brings

together the site agent, job surveyor, job buyer and construction manager. The estimator for the contract should also be present to obtain feedback on the estimate, but in practice (due to a heavy workload) they are rarely present at such meetings. A standard agenda is followed during the meeting which is chaired for its major part by the construction manager. The meetings should be open and honest with the reviews of finances and programme v progress being realistic and accurate.

The headings given on the standard agenda should be broad and give direction rather than specifying in detail the matters to be discussed. The nature of the contract and the smoothness with which work is progressing determine the level of detail appropriate for each section heading. The paragraphs below highlight common subject matters and typical discussion topics.

A Programme

The site agent reviews the progress v programme and agreement is reached as to the actual progress on principal activities and on the contract as a whole. Ways of increasing the productivity on the site are discussed especially in respect of any problem trades. Improved methods of construction and means of effectively sequencing the trades are discussed.

B Financial

The surveyor reports briefly on the current Cost Value Comparison and reviews the overall financial status of the contract. Adverse trends in the profitability of the contract would be discussed and means of reversing the trends examined. A detailed breakdown of cost and value for the cost centres may be given if the contract is not meeting its forecasted profit.

C Buying

The buyer may chair this part of the meeting during which the buying summary schedule is reviewed and the materials requirements of the following month highlighted. The Buyer explains his placing orders previously requisitioned and may on occasions run through the materials profit/loss statement which lists for each material order placed to date the expected profit. The agent and manager review the following month's requirements and note is made of the materials needing to be requisitioned. The position of bulk orders is examined; the amount of materials on site, any loss, wastage, surpluses, etc. Appropriate measures would be agreed to rectify any problems. A longer term look at requirements is taken and any information needed for materials required later in the contract may be requested from the buyer to be brought to the next meeting.

D Nominated Subcontractors

A brief review of their progress, 'manning levels', information requirements etc., contributions are from all parties as appropriate.

E Direct Subcontractors

As D with greater emphasis on manning levels, performance and their financial position. Also there is a review of the subcontracts still to be placed.

F Labour Only Subcontractors

The meeting will consider the following questions: What work are they doing?, Is it being done effectively?, Can we reduce the number of men on site?, What work is there to be done in the coming months?, What manning levels are needed?.

G Direct Labour
(As F above).

H Plant Requirements and Loss

The meeting will consider and examine: An examination of Plant currently on site and whether it can be 'off hired'. A review of lost, stolen or damaged plant and a look at future plant requirements. A listing of Plant to be requisitioned.

JOB REVIEW MEETING FILES
Entry Type: Data Element/Local File

LABOUR ALLOCATION RECORDS
Entry Type: Data Element/Local File

LABOUR COST (+BONUSES)
Entry Type: Data Element/Local File

LABOUR RECORDS
Entry Type: Data Element/Local File

LABOUR TIME SHEETS
Entry Type: Data Element/Local File

LATE PAYMENT CHASER
Entry Type: Data Element

LATE QUOTATIONS
Entry Type: Data Element/Local File

LETTER OF INTENT/ CONTRACT DOCUMENT
Entry Type: Data Element

LIAISON WITH BUYER
Entry Type: Process

Specification:

This process deals with the following data: copy tender enquiry abstracts, subcontract quotations, tender enquiry abstracts, s/c quotes comments, subcontract comments, subcontract quotations and completed tender enquiry abstracts.

A photocopy of each tender abstract is passed to the chief buyer who furnishes the enquiry clerk with additional names of subcontractors and suppliers to whom enquiries/invitations to tender should be dispatched (see section 1.4 in DFD No.1).

The estimator and chief buyer liaise together during the comparison stage of analyzing quotations received back from suppliers and subcontractors.

LIAISON WITH ESTIMATOR
Entry Type: Process

Specification:

This process deals with the following data: s/c names, tender documents.

During the preparation of the estimate there is little formal contact between the tender estimator and planner. Informal contact is frequent, however, with the estimating and planning departments. The planner usually seeks from the estimator the names of the subcontractor's tendering for major works which helps in his consideration of construction methods and resource levels.

Visits to site and the architect's office are undertaken together and it gives ample opportunity for sharing and discussing the characteristics of the projects and its problems.

LIAISON WITH MD & AREA MAN
Entry Type: Process

Specification:

This process deals with the following data: complexities/ alternatives, recommendations, draft programme, area manager comments re prog & prelim sched, draft site organisation structure.

Both the managing director and the contracts manager are approached to discuss any aspect of the tender during its preparation. In general, liaison is usually limited due to their constant involvement and commitment to current projects. The managing director is consulted primarily over the organization structure that could be expected should the contract be awarded. Further discussions may be held to clarify methods of construction where several alternatives exist or when the proposed methods are very complex.

The contracts manager is presented with the pre-tender programme and preliminary schedule as soon as they are complete. (For small projects the contracts manager may prepare them himself). He examines and recommends any modifications necessary. Prior to this the manager is usually involved in making policy decisions such as the type of scaffold that should be used, etc. The manager will be fully conversant with the content of the preliminaries schedule before the finalization meeting during which he may press for changes based upon his own assessment of the resources necessary.

LIAISON WITH PLANNER
Entry Type: Process

Specification:

This process deals with the following data: tender documents, s/c names.

During the preparation of the estimate there is little formal contact between the tender estimator and planner. Informal contact is frequent, however, between the estimating and planning departments. The planner usually seeks from the estimator the names of the subcontractors tendering for major works. This helps in his consideration of construction methods and resource levels.

Visits to site and the architect's office are undertaken by the estimator and planner. This gives ample opportunity for sharing and discussing the characteristics of the project and its problems.

MAIN CONTRACT FINAL ACCOUNT FILES
Entry Type: Data Element/Local File

MANAGER'S INPUTS
Entry Type: Data Element

MANAGER'S MINUTES
Entry Type: Data Element

MANAGERS' INPUT
Entry Type: Data Group

Composition:

MANAGERS' PROMPTS + MANAGER'S INPUTS + DIRECTOR & MANAGER VISIT + INTERNAL MEMOS + TELEPHONE CALL

MANAGERS' INPUTS
Entry Type: Data Element

MANAGERS' PROMPTS
Entry Type: Data Element

MANAGING DIRECTOR'S FILE
Entry Type: Data Element

MATERIAL & S/C QUOTATIONS & COMPARISON SHEET
Entry Type: Data Element

MATERIAL INCREASE FILE
Entry Type: Data Element/Local File

MATERIAL ORDERS
Entry Type: Data Element

MATERIAL REQUIREMENT
Entry Type: Data Element

MATERIAL REQUISITION
Entry Type: Data Element

MATERIAL RETURNS
Entry Type: Data Element/Local File

MATERIAL RETURNS ORDERS
Entry Type: Data Element/Local File

MATERIAL RETURNS SHEET
Entry Type: Data Element

MATERIAL,PLANT & TRANSPORT COSTS
Entry Type: Data Element/Local File

MATERIALS IN WORKS
Entry Type: Data Element

MATERIALS MANAGEMENT
Entry Type: Process

This process includes the following sub-processes: initiate procurement, selection of suppliers, develop contracts, complete purchase procedure.

MATERIALS ON SITE
Entry Type: Data Element

A sum for materials on site will be included in the valuation application.

MATERIALS REQUIREMENT
Entry Type: Data Element

MATERIALS SHORTAGE
Entry Type: Data Element

MD'S MINUTE FILES
Entry Type: Data Element/Local File

MEASURED RATES
Entry Type: Data Element

MEASURED WORK
Entry Type: Process

Specification:

This process deals with the following data: subcontract invoices, work section (trade) total, work measurements, measurement records, valuations files.

Generally the bulk of the work contained in the preparation of a valuation occurs under this heading. The contractor's surveyor is required to itemize the elements of work undertaken up to the date of the valuation and to enter a quantity against each. These quantities are then extended by the applicable bill rate to give each item a value. Work section values are then found and carried forward to a summary sheet. Large or complex contracts may necessitate an alternative method of preparation which is less demanding on time. One such method is to split the contract into defined work packages (usually trades) and to calculate the value of each on the basis of the proportion or work completed, i.e. a percentage figure is calculated for each work section which accurately reflects the amount of work done as a proportion of the total work to be done and this percentage is then applied to the corresponding bill value for the appropriate work section.

The surveyor usually works from the previous month's valuation adding in new work items and revising quantities which have changed during the month. Quantities are either measured physically on site or taken off from the relevant drawings. In addition the surveyor has past measurement records, the quantities in the bills of quantities and also the subcontractors interim applications to aid in the preparation of the valuation. To help in this and to co-ordinate the subcontract valuations to the main contract valuation the surveyor requests that subcontractors submit their interim applications in the week prior to main account valuation. This enables him to cross-check the applications against his own measurements and ensure that all items of work on the applications are covered in some way in the main valuation.

MEASUREMENT FILES
Entry Type: Data Element/Local File

MEASUREMENT OF VARIATIONS
Entry Type: Data Element

MEASUREMENT RECORDS
Entry Type: Data Element/Local File

MEETING AGENDA
Entry Type: Data Element

MEETINGS WITH SUB-CONTRACTORS
Entry Type: Process

Specification:

This process deals with the following data: most competitive quotations, competitive quotations & meeting records, records of past performance, contract programme F., negotiations /meetings.

A buyer will usually approach competitive subcontractors to negotiate further reductions in rates. Having extracted the final offers from the subcontractors he would arrange to meet the lowest with the construction manager. The purpose of the meeting would be to discuss such matters as:

i Programme (start date, key milestones and duration)
ii Expected resource requirements
iii Attendances
iv Terms and conditions of subcontract
v Adjustment of rates to incorporate any reductions & discounts

Such meetings may also be attended by contractor's surveyor, but in any event, all contractor's staff should be satisfied that the subcontractor's performance is likely to be satisfactory. When a subcontractor fails to satisfy or when it is difficult to decide between two or more subcontractors, further meetings may occur to which other subcontractors may be invited.

METHOD STATEMENT
Entry Type: Data Element

This document describes in clear and unambigious terms, how the contractor intends to construct a specific building. It will describe both strategy, operational methods, resources needed and identify key milestones.

METHODS /LOGIC
Entry Type: Data Element

MINUTE CORRECTIONS
Entry Type: Data Element

MINUTES
Entry Type: Data Element

MINUTES & ACTION NOTES
Entry Type: Data Element

MINUTES AGENDA & ACTION NOTES
Entry Type: Data Element

MONTHLY CONTRACT REPORT/CVC FILES
Entry Type: Data Element/Local File

MONTHLY COST + COPY INVOICES
Entry Type: Data Element

MONTHLY CVC SUMMARY
Entry Type: Data Element

MONTH'S COPY MATERIAL,PLANT INVOICES
Entry Type: Data Element/Local File

MOST COMPETITIVE QUOTATIONS
Entry Type: Data Element

NEGOTIATE WITH SUPPLIERS
Entry Type: Process

Specification:

This process deals with the following data: revised quotations, competitive pre-tender quotations, competitive quotations, negotiations.

Upon receipt of additional quotations a buyer analyzes and compares them with the pretender quotations. Should he believe the price or the terms could be improved negotiations would take place with the more competitive suppliers. The best possible rates, terms and conditions are sought.

NEGOTIATED SAVINGS & EXPECTED MARGINS ON S/CS
Entry Type: Data Element

NEGOTIATED SAVINGS & EXPECTED MARGINS ON SUBS
Entry Type: Data Element

This data element includes negotiated savings and expected margins on subcontractors, which represents extra "profit" that can be made on a particular trade.

NEGOTIATION
Entry Type: Data Group

Composition:

NEGOTIATION + REVISED QUOTATIONS

Negotiations will take place with mainly domestic subcontractors and suppliers to encourage them to offer a keener price.

NEGOTIATIONS
Entry Type: Data Group

Composition:

NEGOTIATIONS + ADDITIONAL ENQUIRIES

Occasionally at the pre-contract, contract stage, additional quotations are sought from a supplier who may been overlooked at the tender stage or who has recently come to the notice of contract/buying staff.

NEGOTIATIONS /MEETINGS
Entry Type: Data Element

With sub-contractors

NEW/REVISED INSTRUCTIONS
Entry Type: Data Element

NORMAL PAYMENT AUTHORIZATION
Entry Type: Data Element

NOTICE OF AWARD OF CONTRACT
Entry Type: Data Element

NOTICE OF OFF HIRE
Entry Type: Data Element

NOTICE OF PAYABLE INVOICE
Entry Type: Data Element

NOTICE OF PAYABLE INVOICES
Entry Type: Data Element

NOTICE OF S/C REQUIREMENT
Entry Type: Data Element

NOTIFICATION OF URGENT MATERIAL REQUIREMENT
Entry Type: Data Element

NOTIFICATION OF VALUE TO BE CERT & VAL DATE
Entry Type: Data Element

NOTIFY CERTIFYING VALUE & VALUATION DATE
Entry Type: Data Element

NOTIFYING SUBCONTRACTOR OF VALUATION DATES
Entry Type: Process

Specification:

This process deals with the following data: short term programmes /targets, subcontractor record files, valuation dates.

After having agreed the main contract valuation dates with the pqs the surveyor will usually notify both the nominated and specialist trade subcontractors of the agreed dates and request the submission of interim applications in the preceding week. As further dates are agreed or as circumstances force the valuation dates to change the subcontractors will be informed by letter and continually kept up to date.

NSC & NS COPY INVOICES
Entry Type: Data Element

NSC = Nominated Sub-contractor
NS = Nominated Suppliers

NSC'S INVOICES + NS'S INVOICES
Entry Type: Data Element

NSC = Nominated Sub-contractor
NS = Nominated Supplier

OBTAIN AGREEMENT TO SUBCONTRACT
Entry Type: Process

Specification:

This process deals with the following data: acceptance of order, acceptance chasing letter, subcontract order & appendices, acceptance files, s/c order register, subcontract order notification, subcontract order details, site copy order &

appendices, accounts, buying, s/c order & appendix.

Before the subcontract order is dispatched to the subcontractor the buyer normally enter the following details into a subcontract register:

1 Subcontract Order Number
2 Date of Order
3 Subcontractor's Name
4 Trade
5 Date for Return of Acceptance Form
6 Letter sent
7 Set up

Item 5 above refers to an acceptance of order form which is normally included with each order. The acceptance form requests that the subcontractor signs it to show agreement to the terms and conditions of the order and that he returns the form usually within 14 days of receipt. The form also requests that the subcontractor detail the company's tax exemption provisions together with its VAT registration number. Insurance details as listed on the form are also sought.

Return of the acceptance form is monitored by the buyer through daily reference to the register. A standard letter chasing the form is dispatched if it has not been received by the due date. Upon its return the buyer completes a subcontract order notification form. This form is usually sent to the accounts department and details the main characteristics of the order and the tax and VAT information extracted from the acceptance form.

OFFICE F.
Entry Type: Data Element

OFFICE VISIT REPORT
Entry Type: Data Group

Composition:

OFFICE VISIT REPORT + SITE VISIT REPORT

OFFICIAL SITE MEETING MINUTES
Entry Type: Data Element

ONE WEEK PROGRAMME
Entry Type: Data Element/Local File

ORDER DETAILS
Entry Type: Data Element

ORDERS

Entry Type: Data Group

Composition:

MATERIAL ORDERS + EXTRA ORDERS + CALL OFF REQUEST + SIGNED DELIVERY TICKETS + INVOICE PAYMENT/ REJECTION + NEGOTIATIONS

ORGANISE EXTERNAL PRECONTRACT MEETING
Entry Type: Process

Specification:

This process deals with the following data: immediate info requirements, agreement on 1st meeting.

It is the responsibility of the construction manager to arrange the external precontract meeting. This is usually achieved by a telephone call to the architect after notice of the award of the contract has been received. In addition to arranging the date, time and location of the meeting the manager makes mention of the most urgently needed information; setting out details, bending schedules, etc and requests their receipt as soon as possible.

ORGANISING DECISIONS
Entry Type: Data Element/Local File

P.Q.S FILES
Entry Type: Data Element/Local File

P.Q.S = Professional Quantity Surveyor

PARTICULARS OF MAIN CONTRACT
Entry Type: Data Element/Local File

PAYMENT
Entry Type: Data Element

These are the monies received as a result of the architect's instruction

PAYMENT + INVOICE REJECTION
Entry Type: Data Element

PAYMENT QUERIES
Entry Type: Data Group

Composition:

LATE PAYMENT CHASER + ARCHITECT CERTIFICATE

PAYMENTS
Entry Type: Data Group

Composition:

PAYMENTS + FINAL PAYMENT

PERFORMANCE REPORTS
Entry Type: Data Element/Local File

PETTY CASH ACCOUNT
Entry Type: Data Element

PETTY CASH VOUCHER
Entry Type: Data Element/Local File

PHOTOCOPY AND CHECK INVOICE
Entry Type: Process

Specification:

This process deals with the following data: material returns sheet, invoices, arithmetically correct invoices, interim invoice F., accounts contract F..

Suppliers are normally requested to submit all invoices direct to the regional office. Upon receipt, they are usually passed straight to the invoice checking section of the accounts department where they are photocopied. A copy enters a master file for the contract which will contain a copy of all invoices relating to that contract. Another is entered into an interim file which is usually given to a surveyor each month to help check and validate the contract costs. Invoices are also checked arithmetically to ensure they are correct.

PLACE MATERIAL ORDER
Entry Type: Process

Specification:

This process deals with the following data: selected quotation, order details, regional office files, accounts F., site files, copy orders buying F., material orders.

Having selected the supplier a buyer completes an order form which is immediately dispatched to the supplier's office. Each order is numbered and cost coded. The quantity, description and agreed price is stated with the site address and delivery instructions. Each order carries a short statement requesting that invoices be submitted in duplicate, each month stating the full order number. The reverse side of an order usually states general conditions relating to such matters as quality, price, delivery, quantity, risk, acceptance, cancellation, indemnity, liability and payment of accounts. All terms and conditions of the supplier's quotation are normally excluded from the order unless specifically mentioned on the front sheet as being part of the order.

Orders fall into a number of categories depending upon the nature of the materials and the information known at the time of placing the order. These include:

i Specific one off orders which state precisely the requirements.

ii Open ended bulk orders which indicate the magnitude of the supply requirements from the bill of quantities but which require the agent or the buyer to supply further precise requirements at a later date, e.g. steel reinforcement.

iii Bulk orders which state as accurately as possible at the time of ordering the quantities required and from which the agent subsequently calls off deliveries, e.g. bricks and blocks.

iv Covering orders to cover consumables and sundries. This generally takes the form of an order with a local builders merchant to supply general light materials as and when required. This replaces local purchase orders which are not used within the company.

PLACEMENT OF REMAINING SUBCONTRACT ORDERS
Entry Type: Process

Specification:

This process deals with the following data: subcontract requirement, internal review meeting minutes, subcontract requisition, notice of S/C requirement, buying summary schedule F., contract programme F.

After subcontract orders have been placed for the trades which are immediately required on site (see section 2.6.2 in DFD No. 2.6), normally the buyer, construction manager and quantity surveyor should seek to place the remaining subcontracts as soon as possible, usually within three or four months of the start on site. Urgently required follow-on subcontractors may be requisitioned from site by the agent on a general requisition form. This may not be necessary, however, with the first internal review meeting (see section 3.8 in DFD No.3) giving an opportunity for the subcontract requirements to be discussed and priorities for the placing of orders agreed. Monthly meetings provide a good opportunity to review the progress of the placing of subcontract orders throughout the contract.

The buying summary schedule serves as the control document to both the site agent and the buyer. Sheets are used to list out the subcontractors required on the contract. Against each trade on the schedule, the dates by which they are required on site and by which a requisition should be raised are also listed. By using this schedule the agent can be prompted to requisition subcontractors in good time and the buyer can constantly review the needs of the contract.

PLAN/ PROGRAMME EARLY TRADES
Entry Type: Process

Specification:

This process deals with the following data: contract programme F., agreed weekly programmes, eight week & weekly programme.

Section 2.5.1 in DFD No. 2.5 described the upgrading of the pre-tender programme into a master contract programme which becomes the control against which the progress of the works can be measured. In addition, the company uses a rolling eight week programme updated every four weeks to plan in more detail the work to be carried out. This eight week programme drafted in bar chart format onto standard programme sheets is either prepared by the construction manager or site agent depending upon the programming experience of the site agent concerned. If prepared by the agent, the programme is checked by the construction manager before being adopted as the programme to work to.

In addition to the above the site agent is required to produce a weekly programme each Friday which itemizes activity by activity the work to be undertaken during the following week. The weekly programme has to be agreed with the general foreman and the foreman of the principal subcontractors, who between them carry the responsibility of ensuring the work is subsequently completed. The construction manager monitors the weekly programmes during his visit to site to ensure they are drafted and that the work programmed for the week is actually done. The weekly programme sheet is distributed to the foreman of all the principal subcontractors and to the company general foreman.

PLANNING
Entry Type: Data Element/Local File

PLANNING PROCEDURES
Entry Type: Process

Experienced staff will determine at this stage ways in which the project might be built, together with key milestones and likely rates of construction necessary to meet any imposed completion dates.

PLANT DELIVERY TICKET
Entry Type: Data Element

PLANT DEPT COPY ORDER
Entry Type: Data Element/Local File

PLANT DOSSIER
Entry Type: Data Element/Local File

PLANT ENQUIRIES
Entry Type: Data Group

Composition:

ENQUIRIES + NOTICE OF OFF HIRE + PLANT ORDER

This is normally limited to the contractor's method statement.

PLANT INVOICE
Entry Type: Data Group

Composition:

PLANT INVOICES + TERMINATE HIRE NOTES

PLANT INVOICE VERIFICATION
Entry Type: Data Group

Composition:

INTERNAL PLANT INVOICE + EXTERNAL INVOICE VERIFICATION

PLANT INVOICES
Entry Type: Data Element

PLANT MANAGEMENT
Entry Type: Process

This process includes the following sub-processes: plant requirement schedule, select hire company, hire plant accept procedures, completion of plant hire agreement.

This process deals in careful detail with the terms and conditions of hire (and off hire), such as liability for breakdown, damage and maintenance.

PLANT MOVEMENT BOARD
Entry Type: Data Element/Local File

PLANT ORDER
Entry Type: Data Element

This will be made in conjunction with the 'method statement' and the agreed preliminary schedule.

PLANT QUOTATIONS
Entry Type: Data Group

Composition:

PLANT DELIVERY TICKET + INVOICES + HIRE RATES + QUOTATIONS

The information and quotations obtained at the pre-tender stage, are used as the basis for both contracts with plant-hire companies and for agreeing a schedule of requirements, at the construction stage.

PLANT REQUIRE
Entry Type: Data Element

PLANT REQUIREMENT
Entry Type: Data Element

PLANT REQUIREMENT SCHEDULE
Entry Type: Process

Specification:

This process deals with the following data: plant require, plant schedule F., contract programme F., prelims schedule method statement, plant requisition, construction manager's F., site requisitions.

As work proceeds on the contract both mechanical and non-mechanical plant will be required. Plant required for the initial weeks and even months on the contract is hired (or purchased) and transported to site under the procedures outlined in sections 2.5.5 in DFD No.2.5 and 2.6.3 in DFD No. 2.6. The procedures stated below are associated with the hire and off hire of plant throughout the life of the contract. This is normally administered through a company's own plant department.

Generally the first process in the acquiring of plant is for the site agent to complete a plant requisition/plant terminate hire note. To do this the agent has to have become aware of the need for an item (or items) of plant to be delivered to the contract. Various mechanisms will exist to prompt preplanning in this respect, most important being the monthly internal review meeting (see Section 3.8 in DFD No. 3) during which the future plant requirements will be discussed by the site team. These discussions prompt the early requisitioning of plant and ensure that delay due to late requisitioning is minimized if not eradicated. In addition to this meeting the agent will have the contract programme, the preliminaries schedule and the planners' method statement to aid him in forecasting plant requirements and as a last minute reminder the weekly programme completed with a foreman on a friday requires that the plant and labour strengths for the following week be itemized.

The construction managers who regularly visit the site are

able to stand apart from the day to day pressures of site management and take a clearer forward view of the resource needs on a contract. As such they have a large part to play in ensuring the agent requisitions plant in good time.

The plant requisition/plant terminate hire note requires the agent to describe the item of plant required and state the date it is required and the approximate duration of the hire. It is completed in triplicate, the original going to the plant manager and the copies going to the agent's site file and to the construction manger. The manager's copy is subsequently passed to the surveyors for filing in the regional office contract files.

PLANT REQUISITION
Entry Type: Data Element

PLANT RETURNS
Entry Type: Data Element/Local File

PLANT RETURNS ORDERS
Entry Type: Data Element / Local File

PLANT SCHEDULE F.
Entry Type: Data Element/Local File

PLANT-HIRE COMPANIES
Entry Type: Terminator

This terminator composes the following sub-terminators: plant-hire companies, selected external hire company.

PRE-CONSTRUCTION PROCEDURES
Entry Type: Process

If the tender is successful the contractor will create a project team, and will charge them with the task of preparing detailed, method statements, programmes & financial budgets for the project. Sub-contract and materials orders will be placed. The site organisation will be designed and set up.

PRECISE REQUIREMENTS
Entry Type: Data Element

The precise requirements are determined by looking at working drawings and comparing the scope of the work required with pre-tender subcontractor quotations.

PRELIMINARIES
Entry Type: Process

Specification:

This process deals with the following data: preliminaries schedule, preliminaries totals, preliminary records.

This process is concerned with the system that establishes the value of preliminaries at the dates of external valuation.

Preliminaries are usually either time, value, quantity or occurrence related with the majority being time related or having some time component to its value.

Time related items such as staff salaries and site accommodation are normally valued by dividing the total item value by the expected duration of its operation to get a value per time period (week, or month as appropriate) and multiplying this figure by the duration of its operation to date. (The resulting figure does not always accurately reflect the true expenditure to date but normally suffices for an external valuation).

Value related items such as insurances are most frequently valued in like manner to time related items for the purposes of external valuations.

The value of quantity related items such as concrete test cubes is calculated by finding the product of the number of tests to date and the appropriate bill rate for that item. Occurrence related items such as cleaning or specific protection items are included in the valuation as and when they occur using the lump sum figure in the bills of quantities.

On a typical project over 90 per cent of the total preliminary value is contained in the following six items and as such should be afforded most time and care in preparation.

1 Staff
2 Plant (mech & non-mech)
3 Scaffold & access platforms
4 Site accommodation & warfare
5 Electric and power & other services
6 Cleaning & site tidying

PRELIMINARIES SCHEDULE
Entry Type: Data Element/Local File

PRELIMINARIES TOTALS
Entry Type: Data Element

PRELIMINARY RECORDS
Entry Type: Data Element/Local File

PRELIMINARY SCHEDULE
Entry Type: Data Element

The preliminary schedule will be compiled from information derived from two sources:

1. The method statement, which will determine the nature and quantity of resources (human, plant & equipment) needed on the project.

2. The pre-tender programme, which will determine the length of time each resource will be needed on site.

PRELIMS SCHEDULE METHOD STATEMENT
Entry Type: Data Element/Local File

PREPARATIONS FOR JOB REVIEW MEETING
Entry Type: Process

Specification:

This process deals with the following data: plant schedule F., daily diary sheets, buying summary schedule F., valuation files, buying summary, agenda, current CVC report, s/c lab, plant & mat reports, programme v progress report, progress measurements, contract programme F.

During the monthly site review meeting each party present normally has some input to make and as such a certain degree of preparation is required. The agent has to check the actual progress on principal activities and be able to relate and compare it to the expected (programmed) progress. The monthly progress review sheet may be used to present this information. The agent is required to be familiar with the position of subcontracts, labour, materials and plant and may summarise these matters into a brief report to aid him in the ensuing discussions. The performance of nominated and domestic subcontractors is usually evident and does not require detailed checking. The minutes of subcontractor progress meetings may be referred to for relevant information, as will the labour allocation sheets which record the tasks the direct labour has been undertaking.

The construction manager chairs the majority of the meeting and is required to issue the meeting agenda. The company will normally have a standard agenda which is used on all contracts.

The job surveyor is required to take the current cost value comparison to the meeting and to present a brief overview of the financial position on the contract. A degree of preparation may be necessary.

The job buyer who takes the buying summary schedule and occasionally the materials and subcontract profit and loss statements to the meeting may also be required to gather information requested at a previous meeting. Typical examples of the information requested are, the alternative materials available to meet a particular specifications, the probable suppliers of such materials and the likely delivery periods that could be expected. This information may need to be gained from outside sources or may be obtained through reference to the buying department's products data files.

PREPARE & ATTACH COST BACK UP INFO
Entry Type: Process

Specification:

This process deals with the following data: cost report, month's copy material, plant invoices, monthly cost + copy invoices, cost print outs.

A monthly cost report will usually consist of the following:

- i Contract Report Summary Sheet
- ii Final Reserve Entry Document
- iii Photocopies of any material and Plant Return Sheets which contain accruals
- iv Contract Cost summary
- v Weekly Contract Activity Reports
- vi Contract Labour Detail
- vii Summary of Internal Plant Charges
- viii Photocopies of all materials and external plant invoices received and costed during the month.

The copy invoices will usually be taken from the interim contract invoice files which are filled during the invoice verification process. All the sheets and invoices are stapled together to form the report which is passed to the contractor's surveyor as soon as it is completed.

PREPARE CONTRACTOR'S REPORT
Entry Type: Process

Specification:

This process deals with the following data: information required file, actual progress measurements, information requirements, details of subcontractor performance, contract programme F., subcontractor performance files, information required schedule, contractor's report.

The contractor's report is usually tabled as an item on the agenda and is prepared by the site agent in conjunction with his construction manager. The three principal sections to the report are typically:

- a Progress Review
- b Labour Report
- c Review of Nominated Subcontractors

The content and preparation of each section is given below.

a. Progress Review

Prepared on site by the agent this initial section is usually split into four subsections:

i General
ii Detailed
iii Delays Encountered
iv Delays Foreseen

i The general review serves as an introduction to the report and highlights the overall progress to date on the contract and the work done in the previous month. An overall assessment of the progress of the works in relation to the programmed progress is given.

ii The detailed review is set out on monthly progress review sheets. Each activity programmed to have started prior to the meeting is given a brief description and marked up in terms of the percentage of work actually completed and the forecast percentage of work that will be completed at the end of the next month. The last column of the review sheet is used to briefly remark on and draw attention to those activities on which actual progress is notably ahead or behind the programme progress. In addition, at the end of the detailed report the reasons for these variances from the programme may be expanded in greater detail.

iii The third section entitled 'delays encountered', is used by the agent to record what in the view of the company are the reasons for any delay on the contract. It may simply be a statement of the number of days lost due to inclement weather or may itemize, more specifically, reasons for delay such as late information from the design team, unavailable materials, problems with nominated subcontractors etc.

iv The fourth section 'delays foreseen', is used to refer to problems which can be seen arising in the future and which may not be avoidable. Again typical examples of this include the consequences of the late issue of detailed design information or the late nomination of nominated subcontractors, etc.

b. Labour Report

This section of the report involves the agent in summarizing the recorded labour strengths of both the main contractor and the principal subcontractors. Increases or decreases from the previous month are highlighted and the reasons for abnormal variances if any would be given.

c. Review of Nominated Subcontractors

For each nominated subcontractor currently working on the contract a mini report is prepared showing their progress against programme, labour and material availabilities and information requirements. The information used in these reports is gleaned from subcontractor progress meetings held the day or the morning before the main progress meeting. In addition to the preparation of the above report the agent and construction manager also prepare information required schedules. The schedules detail the information and nominations still outstanding together with the dates each is required. A list of the architect's instructions and verbal instructions still awaiting covering variation orders is also prepared for presentation at the meeting.

PREPARE INTERNAL COSTS
Entry Type: Process

For full explanation see DFD No.3.4.2

PREPARE INTERNAL VALUATION
Entry Type: Process

Specification:

This process deals with the following data: external valuation, measurement records, analytical bills of quantities, internal valuation, value to be certified, P.Q.S files.

After having agreed the 'external valuation' with the P.Q.S the company surveyor is normally required to prepare an 'internal valuation'. The 'internal valuation' represents the 'true value' to the company and is prepared by either adjusting the external valuation or by drafting a completely new valuation in an analytical form. In practice the former is most frequently used with an analytical valuation only being prepared for the very large and important contracts.

To prepare the internal valuation from the external valuation the surveyor must adjust for:

1 Overvaluation - Valuation after cost closing date
- Weighted items in external valuation
- Overmeasure in valuation
- Preliminaries

2 Undervaluation - Valuation before cost closing date

3 Provisions - Unprofitable future work

The surveyors usually make no reserves for 'making good' defects after the contract has finished and for the effects of known delays in the progress of the works (eg possible liquidated damages) If this is the case, the extent and value of such items are considered at the regional level by the directors when the monthly CVC reports for all current contracts have been received.

When an analytical valuation is required (this will normally be decided between the surveying director and the contract surveyor) the surveyor will still have to make the adjustments outlined above. After having found this adjusted true value the surveyor through reference to the analytical bill of quantities has to break it down into the following elements:

1 Labour - This includes general labour, contract staff, labour only subcontractors and all accompanying bonuses and expenses.

2 Materials - Includes, direct, nominated and manufactured suppliers.

3 Plant - Includes all mechanical and non-mechanical plant and any transport.

4 Direct Subcontractors

5 Nominated Subcontractors

6 Sundries - Includes insurances, bonds, services etc.

The surveyor will have to break the valuation down into the above elements.

PRESENTATION OF ARCHITECT CERTIFICATE
Entry Type: Process

Specification:

This process deals with the following data: architect certificate, certificate date, certified value, surveying directors a.c. files, payment, external valuation, surveyors' files.

The company will usually request that an architect's certificates be sent directly to the appropriate regional office. Upon receipt of such a certificate, the surveying director will normally draft a letter acknowledging the certificate value and requesting payment. The letter will then be posted to the client with the appropriate copy of the certificate the same day as the certificate arrives in the office.

Upon receipt of the cheque from the client the surveying director will normally mark up the date of its arrival and value on the 'forecast payments sheet' and the 'contract turnover forecast sheet' The cheque should be banked as soon as possible and the regional accountant notified of its value and arrival.

PRE-TENDER PROCEDURES
Entry Type: Process

'Pre-tender procedures' include all activities associated with tendering. This includes method statements, planning, and obtaining all prices for materials, plant, components, subcontract work and preliminaries.

PRE-TENDER QUOTATIONS

Entry Type: Data Element/Local File

PRE-TENDER QUOTES
Entry Type: Data Element/Local File

PREVIOUS CVC'S FILE
Entry Type: Data Element/Local File

PREVIOUS VALUATIONS
Entry Type: Data Element/Local File

PRICE BILLS OF QUANTITIES
Entry Type: Process

Specification:

This process deals with the following data: unit rates & quotation, build up & extensions, verified build ups & extensions, measured rates.

From the basis of the calculated unit rates for labour, plant and materials, the estimator builds up measured rates against the bill of quantity items. An analytical bill showing labour, plant and material sub-totals is also completed. All workings, build-ups and extensions are passed to the comptometer operators for arithmetical checking.

The provisional and P.C. sums are examined and associated attendances and mark ups added in. A provisional sum and P.C. sum analysis sheet is completed and enables the estimator to calculate the cumulative totals of provisional sums, nominated suppliers and nominated subcontractors. From these totals the discounts are calculated and deducted from the overall total leaving a net total value of all provisional and P.C. sums.

The rates from subcontractors' tenders selected for inclusion in the estimate are transferred to the relevant Bill items after attendances have been added in. A summary sheet of the subcontractors selected for inclusion in the estimate is completed. This subcontract comparison sheet records for each subcontractor the amount of the quotation, the amount in the bills of quantities, the discount rate, the values of savings due to discount and the difference in rates and whether the quotation is fixed price or fluctuating.

PRICE BOOKS
Entry Type: Data Element/Local File

PRICED DAYWORKS
Entry Type: Data Element

PRICED PRELIMINARIES
Entry Type: Data Element

PRICING THE PRELIMINARIES SCHEDULE
Entry Type: Process

Specification:

This process deals with the following data: preliminary schedule, quotations, build up & extensions & work sheet, priced preliminaries.

The resourced preliminaries schedule is priced by the estimator. Quotations obtained by the planner and passed to the estimator form the basis of the rates used for scaffold, major plant and tower crane requirements. Each resourced item is priced although should the estimator consider the allocation to be over-generous or not necessary the item will be reduced or lightly crossed for discussion at the finalization meeting.

The value of the preliminaries is summarized and priced at cost. The priced schedule, together with any workings, etc, are handed to the comptometer operators for checking.

PRIME COST & PROVISIONAL SUMS
Entry Type: Process

Specification:

This process deals with the following data: P.Q.S files, NSc & NS copy invoices, NSc's invoices + NS's invoices, quotations files, section totals.

Prime Cost items are traditionally used in a contract where sections of work or selected materials are to be done or supplied by nominated subcontractors and suppliers. These nominated subcontractors and suppliers are given the dates of external valuations at intervals during the project and their invoices/interim applications for payment are requested to be submitted at least one week before the valuation date. These invoices and applications form the basis of the content of this section of the valuation and copies will be forwarded to the pqs as soon as they are received, thus allowing his prior consideration of them. (The pqs may need to consult other members of the client's team, the clerk of works, the mechanical engineer, etc, to get their assessment of the true value of the work included in these applications).

Provisional sums are included in the contract to cover the expenditure on items that could not be fully specified at the design stage. As these items are subsequently detailed during the construction of the project, the site surveyor will supply rates and quotes to the pqs based on his calculations and when the items are built into the project these quotations (if accepted) will form the basis for calculating the amount to be included in the valuation.

PRINCIPAL ACTIVITIES
Entry Type: Data Element

As a result of the resource exercise, principal activities will be identified, in terms of volume of work, difficulty of supply, or construction, or critical in terms of their influence on either future activities or the overall completion time.

PRINCIPAL S/C'S
Entry Type: Terminator

Comment: Subset of 'Subcontractors' terminator

PROCESS INTERIM APPLICATIONS
Entry Type: Process

This process includes: notifying subcontractor of valuation dates, verify interim application, draft payment authorization, check payment authorization.

PROCURE INITIAL RESOURCE REQUIREMENT
Entry Type: Process

PROCURE MATERIALS
Entry Type: Process

Specification:

This process deals with the following data: requisitions, buying summary schedule, subcontract quotations & comparison sheet, material orders, negotiation, buyer & regional office copy F., project information, contract document F.

After having received the pre-tender quotations, a copy of the tender file and an analytical bill of quantities during the handover meeting, the buyer immediately begins to analyze and procure the initial material requirements.

A typical procedure would be:

1 Analyze pre-tender quotations and comparisons sheets.

2 Check full requirements from boq, spec, etc and from the appropriate requisition if received.

3 Negotiate revised rates with competitive suppliers.

4 Analyze final quotations and provisionally select supplier.

5 Liaise over bulk orders with construction manager and site agent re-prospective suppliers and possible need for visit to supplier's yard or works.

6 Raise order upon acceptable visit to supplier.

7 Place order.

Preference should always be given to suppliers who submitted quotations at the pre-tender stage and hence additional quotations are seldom sought. A fuller description of the procurement procedures for materials is given in section 3.5.1.2 in DFD No 3.5.1.

PROCURE SUBCONTRACTORS
Entry Type: Process

Specification:

This process deals with the following data: contract document F, negotiation, subcontract order, accepting of order, buyer & regional office copy F., project information, detailed /updated info.

At the same time as he is placing material orders, the buyer is also involved with the negotiation and placing of subcontract orders. Working from the pre-tender quotations and analysis sheets the buyer renegotiates with the most competitive subcontractors to obtain favourable rates, terms and conditions. The construction manager and the contract quantity surveyor meet with the subcontractor to discuss their approach to the subcontract works and to ensure they realise resource and programme commitments. After meeting the most competitive subcontractors, the buyer, construction manager and quantity surveyors come to a joint selection decision. In situations where the choice of subcontractor is not obvious or the parties are not in agreement, the construction manager retains the final say. Once a selection decision has been taken the buyer drafts the subcontract order which is dispatched as soon as possible. The buyer monitors the date of receipt of the acceptance of order form which the subcontractor is required to return within 14 days of receipt of the order and upon its receipt notifies the accounts department who proceeds to set up the subcontractor's name on the computer to enable payments to be processed and subsequently dispatched. A fuller description of these subcontractor procurement procedures is given in section 3.5.2.2 on DFD No 3.5.2.

PROCURE COST CENTRE COSTS
Entry Type: Process

Specification:

This process deals with the following data: material,plant & transport costs, subcontract payment, labour cost (+bonuses), sundry costs, petty cash account, site supervisor's claim form, schedule of cost dates F, costs, accruals labour costs.

Each contract within the region of a typical work is costed under the following cost centre headings:

Cost Centre Code	Cost Centre Description
01	Direct Labour
02	Labour Expenses and other Allowances
03	Labour Incentive
04	Labour only Subcontractors
05	Direct Material Purchases
06	Nominated Suppliers
07	Manufacturing Suppliers
08	Plant and transport
09	Direct Subcontractors
10	Nominated Subcontractors
11	Petty Cash
12	Insurances
13	Others

At the beginning of each contract a contractor's surveyor forwards a schedule of 'Cost Dates' to the accounts department to inform them of the week ending dates for which monthly cost reports are required. Several cost reports are required each week from a department and it is usual for the reports to be available to the surveyors toward the end of the next working week. Normally costing is computerized.

Prior to running off printouts highlighting the month's costs the costing personnel will usually check that all current cost information relating to the contracts in question is entered into the computer. Weekly contract activity reports are then accessed from the computer and reveal the details of any transactions occurring during the four or five weeks of the month. These transactions are listed under the appropriate cost headings, reference, the transaction date and supplier's name. The activity reports record all costs other than internal plant and direct labour.

A printout of the monthly internal plant and transport charges is obtained and schedules each item by description, quantity, rate per week and the associated month's charge. The month's total cost is listed under the extended charges.

A type of weekly printout which reveals the remainder of the contract costs is the contract labour sheet. this itemises the costs associated with the remuneration of all directly employed site personnel. These printouts include the true costs resulting from the latter weeks of the previous month.

Having obtained the above cost printouts and a summary of the contract cost, it is possible to run off from the computer and list three cost totals against each cost centre description. The three cost totals represent the cost this month, the cost this year and the cost to date of each cost centre.

PRODUCE METHOD STATEMENT
Entry Type: Process

Specification:

This process deals with the following data: selected methods, recommendations, methods /logic, method statement.

As activities are resourced and optimum solutions calculated, they are entered onto a method statement sheet. Each activity is numbered and given a brief description. (The numbering is not in precise order of construction but does broadly follow the logic of operations). The plant requirements for each activity are also listed.

The average output and quantity of work in each activity, together with resulting durations are listed under the appropriate headings. A note is made of the method sequence (dependant and related activities) together with any remarks considered worthy of mention (such as assumptions made to deduce output levels).

PRODUCE PRELIMINARY SCHEDULE
Entry Type: Process

Specification:

This process deals with the following data: programmed durations, quotations, preliminary schedule, supporting quotations, pre-tender programme statement, draft site organization structure, area manager comments re prog & prelim sched, draft programme, resource data for project strategy.

The standard company preliminaries schedule is filled in by the planner upon completion of the pre-tender programme. This comprehensive list prompts consideration of all the possible project overheads not attributable to measured items. The principal components included in the schedule are:

Summary sheet; Accommodation; Supervision; Site set up; Temporary services; Temporary site electric; Mechanical plant; Cranage; Scaffolding; Protection and sundries; Maintain and tidy; Transport; Special conditions.

The planner may meet with the tender estimator, chief estimator and managing director to discuss informally the proposed site organization structure. The remaining sections, however, will be allocated resources as the planner believes appropriate. The schedule is not priced by the planner but he ensures quotations are acquired when necessary. Should further quotations be warranted at this stage of the planning procedures they are usually obtained over the telephone.

The pre-tender programme and preliminaries schedule are shown to the contracts manager involved with the tender for comment and discussion.

PRODUCE REMITTANCE ADVICE AND CHEQUE
Entry Type: Process

Specification:

This process deals with the following data: notice of payable invoices, remittance advice & cheque.

When an invoice becomes payable the computer system is used to 'run-off' remittance advices and cheques. The cheques are attached to the advice slip and passed to the regional accountant for authorization.

PRODUCT DATA FILES
Entry Type: Data Element / Local File

PROFESSIONALS' INPUT
Entry Type: Data Group

Composition:
SURVEYOR PROMPTS + PROFESSIONALS' INPUTS + QS INPUTS + BUYER'S INPUTS

PROFESSIONALS' INPUTS
Entry Type:Data Element

PROFIT/LOSS STATEMENT F.
Entry Type: Data Element / Local File

PROGRAMME V PROGRESS REPORT
Entry Type: Data Element

PROGRAMMED DATES OF COMMENCEMENT /COMPLETION
Entry Type: Data Element / Local File

PROGRAMMED DURATIONS
Entry Type: Data Element

PROGRESS MEASUREMENT F.
Entry Type: Data Element / Local File

PROGRESS MEASUREMENTS
Entry Type: Data Element

PROGRESS STATEMENTS
Entry Type: Data Group

Composition:
INVOICE DETAILS + WORK MEASUREMENTS + ACTUAL PROGRESS MEASUREMENTS + PROGRESS MEASUREMENTS + NEW/REVISED INSTRUCTIONS

Project: .... W/ / Trade: .... Sheet No: ....

Company													
Item	Quantity	Rate	Sub Total	Rate	Sub Total	Rate	Sub Total	Rate	Sub Total	Rate	Sub Total	Rate	Sub Total

Fig. D2. Quotation analysis sheet

PROGRESS TARGET
Entry Type: Data Element

PROJECT INFORMATION
Entry Type: Data Group

Composition:
PROJECT INFORMATION + INFORMATION REQUIREMENTS + DETAILS OF SUBCONTRACTOR PERFORMANCE

PROJECT INFORMATION PROCEDURES
Entry Type: Process

Comments:

This process deals with project information and includes the following systems:

1. The system for the analysis and distribution of incoming project information.

2. The system for determining what action should be taken on points raised in incoming mail.

3. The system for drafting and dispatching letters and replies to incoming letters.

4. The system for acknowledging architect's instructions.

5. The system for dealing with information received on site.

6. The system for the acknowledgement of architect's instructions on site.

PROMPT TO OFF HIRE
Entry Type: Data Element

Q.S DIRECTOR FILES
Entry Type: Data Element / Local File

QS INPUTS
Entry Type: Data Element

QUALITY CHECK
Entry Type: Data Element

QUOTATIONS
Entry Type: Data Element

QUOTATIONS & COMPARISON SHEETS

Entry Type: Data Element

Comments: An example of this is illustrated in Fig. D3.

QUOTATIONS FILES
Entry Type: Data Element / Local File

RAISE & DISPATCH PAYMENTS
Entry Type: Process

Specification:

This process deals with the following data: normal payment authorization, 'one off' payment authorization, payments, subcontract cost ledger.

The payment details relevant to each subcontractor are usually taken from an authorization form and entered into the computer. The computer system should both cost the relevant contract against the appropriate cost heading and produce a schedule showing the dates of payments to be made. Payment remittances and cheques should be called up on the due date and signed by the regional accountant. The cheques are countersigned by either the surveyor and managing directors or the senior construction manager before being dispatched direct to the subcontractor.

RAISE ADDENDUM ORDERS
Entry Type: Process

Specification:

This process will deal with the following data: extra orders, regional office files, accounts, site F., copy addendum orders buying, supplementary requirement to existing order.

On occasions the buying department will receive requisitions which either vary the content of an existing order or supplement it in some way. Addendum orders are raised which are distributed to the supplier to supplement the initial order. A buyer uses supplementary appendix sheets to communicate such changes to the supplier even though they are more specifically used to appendix matters relating to subcontract orders. In situations when the content of an order has been substantially altered by a variation to the contract, a buyer would usually renegotiate with the supplier and raise a completely new order to replace the original.

RECOMMENDATIONS
Entry Type: Data Element

Comments: This element outlines senior management guidance on key issues.

RECONCILE BULK MATERIALS
Entry Type: Process

Specification:

This process deals with the following data: materials shortage, bulk materials reconciliation files, materials on site, materials in works, delivery details.

Although generally no specific forms are available for such purposes a site engineer usually monitors the delivery, stock levels and wastage of bulk materials. Periodic reconciliations typically monthly, calculate the materials delivered, the amount used in the works and the resulting wastage or loss. An engineer in performing these duties is able to give notice of material shortages on site to the agent who can arrange further deliveries. He is also able to check that wastage or losses remain within acceptable limits and do not exceed the estimator's allowances. Feedback to an estimator of the actual wastage levels rarely takes place, the possible exception to this being when the estimator manages to attend one of the monthly internal review meetings.

RECORD DELIVERY
Entry Type: Process

Specification:

This process deals with the following data: in tray (delivery tickets), delivery details, site MRS sheets, material returns sheet.

A materials return sheet would be completed by the general foreman or the site agent weekly. The sheet will normally record details of all deliveries within any week up until the Saturday. Upon completion, the sheet is submitted direct to the accounts department with one copy being retained on site. Part of the form is completed on site, the remainder can be used by the accounts department for costing purpose.

RECORDS OF MEETINGS
Entry Type: Data Element / Local File

RECORDS OF PAST PERFORMANCE
Entry Type: Data Element / Local File

REGIONAL OFFICE
Entry Type: Terminator

Comment: This is a sub-set of 'Head Office' terminator.

REGIONAL OFFICE (QS)
Entry Type: Data Element / Local File

REGIONAL OFFICE FILE
Entry Type: Data Element / Local File

REGIONAL OFFICE FILES
Entry Type: Data Element / Local File

REGIONAL OFFICE MEETING MINUTES FILE
Entry Type: Data Element / Local File

REMINDER TO REQUISITION OF MATERIALS
Entry Type: Data Element

REMINDER TO REQUISITION OF PLANT
Entry Type: Data Element

REMINDER TO SUBCONTRACTORS
Entry Type: Data Element

REMITTANCE ADVICE & CHEQUE
Entry Type: Data Element

REQUEST TO SUPPLY SHEETS
Entry Type: Data Element / Local File

REQUIRING REPLIES
Entry Type: Data Element

REQUISITION INITIAL RESOURCES
Entry Type: Process

Specification:

This process deals with the following data: buying summary schedule, initial activities' dates, detailed /updated info, requisitions.

The site agent completes requisitions for materials and plant with the construction manager raising subcontract requisitions. The requisitions are drafted and sent to the relevant departments as soon as possible but a common problem during this phase of the contract is a lack of detailed and updated information from the design team. As such the requisitioning of certain critical materials (for example, reinforcement) has to be held back until after the external precontract meeting during which it is usual to be supplied with the necessary drawings and schedules.

REQUISITION TO PLANT HIRE CO
Entry Type: Data Group

Composition:
PLANT ENQUIRIES + PLANT ORDER + AGREED HIRE RATES + NEGOTIATION

REQUISITION TO SUPPLIER
Entry Type: Data Group

Composition:
ORDERS + NEGOTIATION + INSTRUCTIONS TO SUPPLIER

REQUISITION/ TERMINATION F.
Entry Type: Data Element / Local File

REQUISITIONS
Entry Type: Data Element

RESOURCE DATA FOR PROJECT STRATEGY
Entry Type: Data Element

This element includes resource data, site layout plan, method statement, pre-tender programme & programme alternatives.

RESOURCE EXERCISES
Entry Type: Process

Specification:

This process deals with the following data: bulk quantities, temporary works details, s/c names, principal activities, selected methods, complexities/ alternatives.

Having abstracted the bulk quantities for each principal activity the planner proceeds to resource them in terms of their labour and plant content. Using his own output constants for labour and plant, activity durations are calculated initially on the basis of the most sensible and obvious resource strengths. All possible methods of construction are considered and resourced and gang sizes and plant capacities varied to appraise alternative solutions. The optimum solutions are sought for each activity and the planner may liaise with the managing director over the more complex decisions where viable alternatives exist.

RESOURCE FORECAST
Entry Type: Data Element

RESOURCE MANAGEMENT
Entry Type: Process

RESOURCE REQUIREMENT
Entry Type: Data Element

RESOURCE REQUIREMENT (FROM 3.2)
Entry Type: Data Group

Composition:
PLANT REQUIRE + PROMPT TO OFF HIRE + SUBCONTRACT REQUIREMENT + MATERIAL REQUIREMENT + NOTIFICATION OF URGENT MATERIAL REQUIREMENT

RESOURCE REQUIREMENT (TO 3.5)
Entry Type: Data Group

Composition:
MATERIALS REQUIREMENT + S/C REQUIREMENT + PLANT REQUIREMENT + INFORMATION REQUIREMENT

REVIEW & UPGRADE PROGRAMME
Entry Type: Process

Specification:

This process deals with the following data: tender file, staff appointments, activity dates, initial activity's dates, contact programme.

Once the contract has been secured the contractor's construction manager or project manager (on a large contract), confirmed as the manager to supervise the project, obtains the pre-tender programme from the planner together with the preliminaries schedule and method statements. The programme is analyzed by the manager who has the option to either expand or redraft it into a contract programme. As soon as the site agent becomes available he too is brought into the process of finalizing the programme and with the manager produces a programme which suits their preferred methods of working. It is usually not possible to complete the programme at this stage due to a lack of detailed and updated information from the architect and engineers. The programme is taken to the external precontract meeting but normally requires further redrafting before it is issued as the contract programme. Unless otherwise specified the contract programme is presented as a simple bar chart which is on a typical programme sheet and includes references to the dates by which detailed drawings/schedules are required for each activity together with the dates by which nominated subcontractors must be specified.

REVIEW LATE QUOTATIONS
Entry Type: Process

Specification:

This process deals with the following data: trade files,

selected quotations, competitive late quotations, late quotations.

It is common for further subcontractor and supplier quotations to be received after the stated closing date. As and when they arrive they should be analyzed and compared against those already selected for inclusion in the estimate. Should they appear favourable and competitive they are retained by the estimator for possible inclusion into the finalization meeting.

REVIEW MEETING NOTES
Entry Type: Data Element / Local File

REVISED QUOTATIONS
Entry Type: Data Element

S/C INFO
Entry Type: Data Group

Composition:
CASH FLOW FORECASTS + FOREMENS' INPUT + S/C RESOURCE ACCOUNT

This includes information from the sub-contractor to the main contractor largely of a progress, financial and managerial control nature.

S/C LAB, PLANT & MAT REPORTS
Entry Type: Data Element

S/C NAMES
Entry Type: Data Element

Advice from the estimator to contract planning staff.

S/C ORDER & APPENDIX
Entry Type: Data Element

S/C ORDER REGISTER
Entry Type: Data Element / Local File

S/C QUOTES COMMENTS
Entry Type: Data Element

Advice from buyer to estimating team

S/C REQUIREMENT
Entry Type: Data Element

S/C RESOURCE ACCOUNT
Entry Type: Data Element

Composition:
INTERIM APPLICATIONS + FINAL ACCOUNT + DAYWORKS + FOREMEN'S INPUTS TO PROGRAMMES

SAFETY CHECK
Entry Type: Data Element

SAFETY NOTES
Entry Type: Data Group

Composition:
SAFETY NOTES + IMPLEMENTED SAFETY REPORT

SAFETY OFFICERS
Entry Type: Terminator

Comment: This is a sub-set of 'Head Office' terminator.

SAFETY OFFICERS' REPORTS
Entry Type: Data Element / Local File

SAFETY REPORTS
Entry Type: Data Element

SCAFFOLD SCHEDULE PLANT REQUIREMENT
Entry Type: Data Element

SCHEDULE OF COST DATES
Entry Type: Data Element

SCHEDULE OF COST DATES F.
Entry Type: Data Element / Local File

SCHEDULE OF COSTS & ADJUSTMENTS
Entry Type: Data Element

SCHEDULE OF VALUATION & COST DATES (FROM 3.1)
Entry Type: Data Group

Composition:
SHORT TERM PROGRAMMES /TARGETS + SCHEDULE OF VALUATION & COST DATES (FROM 3.1)

SCHEDULE OF VALUATION & COST DATES (TO 3.5)

Entry Type: Data Group

Composition:
SCHEDULE OF COST DATES + VALUATION DATES

SCHEDULE PAY DATES
Entry Type: Process

Specification:

This process deals with the following data: accounts payable schedule, notice of payable invoices.

The computer system would normally print out an accounts payable schedule/non-scheduled analysis of all the invoices to be paid. This schedule should be reviewed by the chief accountant and it should show the dates by which the invoices should be paid to obtain full discount. Invoices which do not carry restrictive payment terms are usually paid at the end of the month in which they are received (if received before the middle of the month) and would not be scheduled in any specific order but grouped with other like invoices. The schedule acts in part as a visual check for invoice duplication but is mainly used to prompt the timely dispatch of appropriate remittances.

SECTION TOTALS
Entry Type: Data Element

SECURE CONTRACT
Entry Type: Process

Specification:

This process deals with the following data: contract programme, letter of intent/ contract document, tender details, senior staff, notice of award of contract, tender information.

It has become increasingly common for a company to be drawn into some form of negotiation before the contract is awarded. These negotiations usually resolve around items such as:

i The clarification and discussion of any qualifications included in the tender.

ii Discussions relating to production alternatives indicated in the submission and their consequential monetary implications (if any).

iii Methods of reducing further the tender price and associated implications in respect of time and quality of construction.

Meetings are usually arranged between the company and consultants to discuss items of major significance. Smaller matters such as the clarification of minor qualifications are dealt with on the telephone and by correspondence.

The choice of personnel to be present at any such meetings

is dependent upon the nature of the discussions. The tender estimator and the directors most frequently attend, the former because of his intimate knowledge of the tender and the latter because commercial decisions may have to be taken. In addition the planner and construction manager involved with the tender preparation may also attend.

It is hoped that such a meeting provides the basis from which the contract would be awarded to the company. A letter of intent or confirmation of the award of the contract would be expected before the company proceeds further with the process of setting up the project.

SELECT HIRE COMPANY
Entry Type: Process

Specification:

This process deals with the following data: plant movement board, plant requisition, plant dossier, site copy order, accounts copy order, plant dept copy order, plant order, hire rates + quotations, enquiries.

The plant manager is usually situated at a regional office or in a separate plant department. Upon receipt of a requisition from site he is responsible for hiring or purchasing a suitable piece of equipment and ensuring it is transported to site on or before the desired date.

The plant department will usually carry a stock of plant which is largely small and medium sized equipment. The plant manager decides whether he can meet the requirement from the stock of company equipment or whether to hire the plant from an external source. In considering this matter the manager also has to decide whether it might be cost effective or beneficial to purchase the item for the company, thus adding it to the current stock of plant. Decisions of this nature depend upon existing stock levels and the condition of plant together with the expected hire duration and assessed future demand. If internal plant is to be used the manager will organize transport to site using the company transport vehicles.

If the manager decides to hire plant externally he will generally have established sources from whom hire rates would be sought. New hire companies may also be considered and given an opportunity to submit a quotation. Selection of a supplier would depend upon many factors including hire rates, reliability, proximity and reliability of service/maintenance crews and past experience. Orders are usually agreed on the telephone and confirmed on a standard company order form. The order gives an item description and states the quantity and agreed hire rate together with any additionally agreed terms and conditions. The back of the order lists standard company conditions relating to such matters as quality, price, delivery, quantity, risk, indemnity and payment of accounts.

The plant manager generally has a plant movement board which monitors the location of internal plant. The board is updated as and when plant is either returned or transferred from one site to another. A plant dossier is often kept at the depot by the manager which contains literature describing all forms of plant on the market. The information is then available for reference and is used by the construction managers in assessing the suitability and capabilities of new and unusual equipment. The dossier is maintained by the plant manager and updated regularly.

SELECT SITE TEAM
Entry Type: Process

Specification:

This process will deal with the following data: staff requirement, senior staff, notice of award of contract, staff appointments.

The directors are likely to become aware of the possibility of a contract being awarded when the bills of quantities are called for submission. At this stage the managing director and the surveying director affirm whether the construction manager involved in the tender preparation will continue with the contract and be responsible for its supervision during the construction period. The manager's presence may be required in the process of securing the contract.

The management structure on any one project is determined by its nature and individual characteristics, (size, complexity, nature of operations, speed of construction, size of workforce, etc.), this will have been considered while formulating the preliminaries schedule during the tender preparations. The selection of appropriate agents, foremen and surveyors is made by the directors in consultation with the construction manager. Particular attention is given to the character of the consultants on the project; architect, pqs, structural engineer, etc and when possible, compatible staff are selected. Other factors affecting the selection of staff are their current availability, location (and the disruption that may be caused by a move), experience of the staff in that type of building works and the compatibility of the staff in terms of being able to create a good site team. Junior staff; assistant surveyors, engineers etc are selected as soon as possible although their early selection is not as essential as that of the senior staff.

SELECT SUPPLIER
Entry Type: Process

Specification:

This process deals with the following data: competitive quotations, works/yard visit notes, selected quotation.

The final selection of a supplier is made after consideration of criteria such as:

i Rates (including discounts)
ii Payment terms and conditions
iii Attendances required
iv Availability of Materials
v Reliability of Supplier
vi Quality of Materials
vii Supplier's ability to meet delivery and quality requirements:

Works capacity
Fleet size
Back up facilities
Locality

When assessing factors such as the supplier's ability to meet delivery and quality standards, it may be necessary to make a visit to the supplier's establishment. A satisfactory feedback from a visit would be necessary before a buyer would proceed to place an order. Visits are only usually made to the suppliers of principal materials such as ready mix concrete and precast cladding panels and typically by the construction managers or agent. Feedback from past supply contracts is obtained by the buyer during his monthly site visit.

SELECTED EXTERNAL HIRE COMPANY
Entry Type: Terminator

Comment: This is a subset of 'Plant-hire companies' terminator

SELECTED METHODS
Entry Type: Data Element

This element consists of the preferred or deemed 'best' construction methods.

SELECTED QUOTATION
Entry Type: Data Element

SELECTED QUOTATIONS
Entry Type: Data Element

These represent normally the lowest quotation in each work or trade group.

SELECTED SUBCONTRACTOR
Entry Type: Terminator

Comment: This is a subset of 'Suppliers' terminator

SELECTED SUPPLIER
Entry Type: Terminator

Comment: This is a subset of 'Suppliers' terminator

SELECTION OF SUPPLIER
Entry Type: Process

This process includes the following sub-processes:
develop a procurement strategy; assess pre-tender quotations; negotiate with suppliers; select suppliers; place material order; complete supplier profit/loss statement; raise addendum orders.

SENIOR STAFF
Entry Type: Data Element

Senior staff involvement with client & his representative details of negotiation dealt with here.

SET UP SUBCONTRACT PAYMENT PROCEDURES
Entry Type: Process

Specification:

This process deals with the following data: subcontract order notification, tax exemption details.

Normally, upon receipt of a subcontract order notification sheet the accounts department feeds the data into the company computer system which sets up the process for subcontract of payments details received from the surveyor during the contract. Without having been set up in this way subcontract payments would be delayed until the subcontract acceptance form had been received back. In certain cases, however, the computer system could be set up without confirmation of the subcontractor's agreement, to enable initial payments to be paid.

SHORT TERM PROGRAMMES /TARGETS
Entry Type: Data Element

SIGNED DELIVERY TICKETS
Entry Type: Data Element

SIGNED FINAL ACCOUNT
Entry Type: Data Element

SITE
Entry Type: Terminator

This terminator composes the following sub-terminators: site foremen, site workforce.

SITE ADMINISTRATION SET UP
Entry Type: Process

Specification:

This process deals with the following data: project information, construction manager's file, site files (agents), regional office file, surveying director's file, managing director's file.

A company will generally have a recommended contract filing system which outlines the expected content and format of both the site and regional office files. The site agent is responsible for setting up and maintaining the site files and likewise the contract surveyor sets up and maintains the regional office files.

The regional office files contain all original correspondence with the site files being made up of copies.

In addition to these two largely comprehensive files the construction managers are likely to retain limited amounts of project information which they carry about with them during travels from site to site.

The quantity surveyor director will typically retain the contract documents for each project in his personal office file, together with the originals of any important contractual certificates such as extension of time certificates, certificates of insurance, performance bonds, etc. Both directors should receive and retain copies of the minutes of all internal review meetings (monthly site meetings on all contracts involving the full site team) and the quantity surveying director should also retain monthly for each current contract the cost value comparison reports.

The normal procedures for handling project information and correspondence received during the construction phase of the contract are dealt with in section 3.6 in DFD 3.

SITE CONFIRMATION OF INSTRUCTION F.
Entry Type: Data Element / Local File

SITE COPY ORDER
Entry Type: Data Element / Local File

SITE COPY ORDER & APPENDICES
Entry Type: Data Element

SITE COPY ORDERS
Entry Type: Data Element / Local File

SITE DETAILS
Entry Type: Data Element / Local File

SITE DIARIES
Entry Type: Data Element / Local File

SITE DIARY
Entry Type: Data Element / Local File

SITE F.
Entry Type: Data Element / Local File

SITE FILE (COPY ORDER)
Entry Type: Data Element / Local File

SITE FILES
Entry Type: Data Element / Local File

SITE FILES (AGENTS)
Entry Type: Data Element / Local File

SITE FOREMEN
Entry Type: Terminator

Comment: This is the subset of 'Site' terminator

SITE INFO MANAGEMENT
Entry Type: Process

Specification:

This process deals with the following data: external correspondence, internal memo files, daily diary files, director & manager visit, daily diary sheets, internal memos, telephone call, formal conversations, informal conversations, dayworks, authorized daywork.

Effective communication between all parties is an essential attributed on any project if it is to be successful. Typically there are four areas of concern:

1 between site and regional office
2 between staff and workforce
3 between the company and its subcontractors & suppliers
4 between the company and the client's consultants.

Internal communications between site and the regional office are many and varied. For example, 'daily diary sheets' submitted to the construction manager and buyer, internal memos to confirm any important internal decisions and more frequently telephone conversations with and the visits to site by the construction manager, surveyors and directors. There should be a realisation within the company of the need to break down any departmental barriers and to develop a teamwork approach to the contract.

DAILY DIARY

Site		Job No.	Date	Weather Report Temperature a.m. .... p.m.	
Labour Supervision	No	Week No	Main Programme Days Behind .... Days Ahead		Short Term Programme Days Behind .... Days Ahead ....
Bricklayers		Description of Work			
Carpenters					
Painters					
Plumbers					
Labourers					
Total					
Sub Contract Lab	No.				
Trade					

1. Materials 2. Plant 3. Information Required 4. Delays 5. Variations 6. Any other items	Advice or assistance required from office
Visitors	Details of Accidents or Thefts

THIS SECTION TO BE COMPLETED ON THE LAST DAY OF WEEK ONLY		Yes	No	Signed
	Has safety inspection been made?			
	Has Buying Summary Schedule been reviewed?			
	Have statutory forms been completed?			
	Have dayworks been sent to office?			....
	Have copies of confirmation of verbal instruction sheets been sent to the office?			Agent/G. Foreman

SPECIAL REMARKS:

Week ended .... 19 ....

Fig. D5. Daily diary sheet

Formal communications with operatives on site should be encouraged to be always up or down the line of responsibility. Informal chats between contract staff and operatives are important, serving to break down any barriers and suspicions that may otherwise exist.

The principal methods of communication between site and external organizations such as subcontractors are normally the telephone and correspondence. When possible all important telephone conversations should be confirmed in writing or at least noted into a diary or correspondence file. The procedures associated with the receiving and sending of letters are dealt with under sections 3.6.1 to 3.6.5 in DFD No 3.6.

SITE INSPECTION REPORT
Entry Type: Data Element/ Local File

SITE LAYOUT PLAN
Entry Type: Data Element / Local File

SITE M.R.S
Entry Type: Data Element / Local File

Comment: M.R.S = Materials Returns Sheet

SITE MANAGEMENT
Entry Type: Process

Comments: This process deals with the transformation of data entering & leaving site system. It involves:

1 The system developing site strategy.
2 The system for site scheduling.
3 Forecasting system for all resource requirements.
4 Site team co-ordination system.
5 The system for site personnel management.
6 The site performance monitoring system.
7 The site information management system.

SITE MEETING MINUTES F.
Entry Type: Data Element / Local File

SITE MRS SHEETS
Entry Type: Data Element / Local File

Comment: MRS = Material Returns Sheet

SITE ORDER F.
Entry Type: Data Element / Local File

SITE ORDER FILES
Entry Type: Data Element / Local File

SITE PERFORMANCE MONITORING
Entry Type: Process

Specification:

This process deals with the following data: drawings, specifications, daily diary, safety officers' reports, quality check, safety check, supervision, implemented safety report, safety reports, priced dayworks, authorized dayworks, dayworks, petty cash voucher, daily diary sheets, labour time sheets, labour allocation records, agreed course of action.

Typically the engineer working under the supervision of the site agent is responsible for setting out the works and providing all levels and bench marks necessary for the work to proceed accurately and smoothly. He is also generally responsible for checking the works as completed for accuracy and tolerance. The site agent and foremen jointly co-ordinate and supervise the work and also inspect the works together to ensure it meets the required specification; quality of finish etc.

The progress on the contract is monitored constantly by the surveyors measuring for valuation and subcontract purposes. In addition the agent has to mark up the master programme weekly which requires him to undertake a weekly measure. This measure is not in great depth but should be sufficient to enable him to monitor whether actual outputs on the principal trades and activities are meeting the planned and expected outputs. In situations where a significant variance between actual and planned outputs is noticed on principal and critical activities, the agent undertakes a more detailed measure from which he is able to consider changes that could be made to resource levels so as to accelerate (or even slow down) production. The agent is required to produce a detailed progress report for the monthly external pre-contract meeting (see section 3.7.2 in DFD No 3.7).

Daily records should be maintained by the foremen of the hours worked by all direct employees on the contract. From these records the agent writes up time sheets for each employee every Friday and submits them to the wages department situated at regional office. In addition daily allocation sheets are completed by the foreman or ganger for the direct labour to record the work and hours undertaken by each. These sheets are signed weekly by the agent and subsequently a copy is given to the surveyor. The labour force on site each day is also recorded onto the 'daily diary sheet', this includes the labourers and tradesmen working for subcontractors.

Any dayworks carried out by the company or by subcontractors on behalf of company is recorded by the general foreman or the agent. It is also picked off the allocation sheets and the subcontractors' day work sheets. The agent is normally responsible for completing the daywork sheets. Four copies are completed and passed to the architect or clerk of works for verification. One copy should be retained by the clerk of works and the remaining three passed to the company surveyor for pricing. Priced copies are dispatched to the architect and pqs for inclusion into the next valuation.

The 'daily diary sheets' are normally completed by either the agent or the general foreman. The contents of the 'diary sheet' can be seen by reference to the example given in Fig. D4. Copies of the sheet are submitted to the buying department and the contracts manager respectively and one copy is retained on site. Although completed daily, the diary sheets are usually submitted weekly to the office, the content of the report is designed to keep the manager and buyer up to date with the work in hand and the requirement of the site in terms of the assistance they can give, (e.g. provide information as to a certain material or item of plant, etc.). The information recorded also aids the preparation of claims should the need arise.

Site safety should be constantly monitored and checked by all members of the site team. The company safety officer will visit the site at approximately four week intervals to conduct safety checks and in addition several of the principal subcontractors are likely to use the services of a safety officer to conduct their own independent checks. The company safety officer should leave two copies of a handwritten report on site which highlights any dangerous and unacceptable aspects of the site and the equipment. A copy of the report is also submitted to the managing director at the regional office for his consideration. The site agent will be required to correct the problems highlighted by the report and upon the completion of this, return one copy of the report back to the safety officer indicating that rectifying action has been taken and stating the date of its implementation.

The site agent normally looks after the site petty cash. A petty cash voucher is sent by the agent to the accounts department weekly indicating the amount spent and the items purchased. The accounts department will cost the expenditure against the contract and replenish the petty cash on site to its original level by dispatching a 'top up' amount of cash direct to site.

Each site agent will normally be asked to complete a monthly movement and claim form each month which indicates the contracts he has been working on and the percentage of his month's salary that should be costed against each. In addition the form could be used to record any overtime worked during the month and this section should be completed with the construction manager who has to authorize the payment of any overtime monies. The form should typically be submitted direct to the accounts department for costing purposes.

SITE PERSONNEL MANAGEMENT
Entry Type: Process

Specification:

This process deals with the following data: bonuses, progress target, performance reports, foreman's inputs.

The site team should place great emphasis on encouraging and motivating the workforce and in developing and maintaining a high morale. A happy site is seen as a good high performing site and to this end the following factors should be seen as important:

1 Welfare
The provision of a good quality canteen providing for both staff and the workforce. The canteen should be subsidized by the company. The provision of adequate clean washing, shower, locker, and toilet facilities should be provided.
2 Safety
To comply with the safety legislation and to keep the site as safe as possible, the company employs its own safety officers who visit site at random. In addition several of the larger subcontractors likely to be used by the company, will have their own visiting safety officers. All staff should wear safety helmets and every effort should be made to ensure the workforce do likewise.

3 Renumeration
To ensure that bonuses reflect fairly and truly the effort and productivity of any direct labour force.

4 Unions
To maintain good relations with union representatives or spokesmen.

5 Communication
To encourage the workforce by being interested and concerned with them as individuals. This involves both the agent and foremen developing friendly relationships with the workforce and maintaining an open door approach encouraging them to voice any problems or dissatisfactions.

6 Discipline
To be seen to take fair and reasonable steps to warn and discipline troublemakers before any penalties or dismissal are considered.

SITE PROGRESS MEETING
Entry Type: Process

Specification:

This process deals with the following data: contractor's report, meeting agenda, project information, consultants' reports, information required schedule, minutes, minute corrections, minutes & action notes.

Usually held once a month, these meetings are generally attended for the company by the site agent, construction manger and occasionally the quantity surveyor. Representing the client it is usual for the architect, clerk of works, pqs and main consultants to be present.

The meeting is most commonly chaired and minuted by the architect and follows either his own agenda or the company standard agenda. The construction manager also takes comprehensive minutes from which to subsequently check the officially recorded minutes issued by the architect. In addition to the issued agenda, an extra agenda form is used to record any significant matters raised during the meeting and also has an 'action by' column to list any action to be taken by all parties after the meeting.

During the meeting the reports shown on the agenda are usually delivered and discussed. Agreement is usually reached over such matters as the degree of description due to inclement weather and the actual progress to date. Information required schedules are handed to the architect together with a list of architects and verbal instructions still requiring covering variation orders. In return the site team may receive information previously requested in the form of drawings, schedules instructions.

SITE PROGRESS MEETING FILE
Entry Type: Data Element / Local File

SITE REQUISITION FILE
Entry Type: Data Element / Local File

SITE REQUISITIONS
Entry Type: Data Element / Local File

SITE SCHEDULING
Entry Type: Process

Specification:

This process deals with the following data: professionals' inputs, managers' inputs, info requirement, resource requirement, one week programme, eight week programme.

The amount of planning and the degree of detail necessary largely depends upon the complexities of the work. It is usual for short term detailed programmes to be produced during the initial weeks on any site and for very detailed programmes to be drafted in the latter weeks. The need for regular short term planning during the rest of the contract will be assessed by the construction manager and the agent together.

SITE STAFF ACCOMMODATION SUPPORT
Entry Type: Process

Specification:

This process deals with the following data: material orders, visit report, negotiation, project information, requisitions.

The contract non-mechanical plant requirements are typically processed in a like manner to the mechanical plant. A form called the 'preliminary forecast of non-mechanical plant requisition' is completed by the site agent and construction manager and sent normally to the plant manager to notify him of the expected site requirements. (This form covers items such as site accommodation, portaloos, hoardings, nameboards and theodolites.) The agent is required to confirm these provisional requirements with requisitions, from which the plant manager processes the selection, ordering and transportation of the items to site.

A company will typically have a standard list of all site stationery upon which at the beginning of a contract the site agent indicates his requirements by marking the quantity desired against each item. The stationery is then gathered and dispatched to site or to the agent by one of the regional office secretarial staff.

SITE SUPERVISOR'S CLAIM FORM
Entry Type: Data Element / Local File

SITE VISIT NOTES
Entry Type: Data Element

SITE VISIT REPORT
Entry Type: Data Element

Comment:
The site visit report will normally include an assessment on the following:
access, services, state of adjacent properties, site office and storage areas, general site conditions, environmental implications for surrounding community.

At the pre-construction stage more detailed information may well be sought. This report is usually provided by the architect, but the contractor may wish to supplement the information by his own visit.

SITE WORKFORCE
Entry Type: Terminator

Comment: This is a subset of 'Site' terminator.

SPECIFICATIONS
Entry Type: Data Element / Local File

STAFF APPOINTMENTS
Entry Type: Data Element

STAFF REQUIREMENT

Entry Type: Data Element

Comment: This element lists the principal skills needed in a chosen site team.

SUB-CONTRACTORS
Entry Type: Terminator

Comment:
The term 'sub-contractors' includes, nominated sub-contractors, and all 'specialist trade contractors' (including the main contractor's own domestic subcontractors). This terminator includes the following sub-terminators: principal sub-contractor, sub-contractor, competitive sub-contractors and selected sub-contractors.

SUBCONTRACT COMMENTS
Entry Type: Data Element

Comment: Advice from buyer.

SUBCONTRACT COMPARISON SHEETS
Entry Type: Data Element / Local File

SUBCONTRACT COST LEDGER
Entry Type: Data Element / Local File

SUBCONTRACT INVOICES
Entry Type: Data Element / Local File

SUBCONTRACT MANAGEMENT
Entry Type: Process

This includes the following sub-processes: placement of remaining sub-contract orders, sub-contractor selection procedures, draft and place sub-contract order, contract procedures, finalization.

SUBCONTRACT ORDER
Entry Type: Data Element

SUBCONTRACT ORDER & APPENDICES
Entry Type: Data Element

SUBCONTRACT ORDER DETAILS
Entry Type: Data Element

SUBCONTRACT ORDER NOTIFICATION

Entry Type: Data Element

SUBCONTRACT ORDERS
Entry Type: Data Element / Local File

SUBCONTRACT PAYMENT
Entry Type: Data Element / Local File

SUBCONTRACT PROFIT/LOSS STATEMENT
Entry Type: Data Element

SUBCONTRACT QUOTATION
Entry Type: Data Element

SUBCONTRACT QUOTATIONS
Entry Type: Data Element

SUBCONTRACT REGISTER
Entry Type: Data Element / Local File

SUBCONTRACT REQUIREMENT
Entry Type: Data Element

SUBCONTRACT REQUISITION
Entry Type: Data Element

SUBCONTRACTOR FILES
Entry Type: Data Element / Local File

SUBCONTRACTOR LIABILITIES
Entry Type: Data Group
Composition:
SUBCONTRACTOR LIABILITIES + PROGRESS MEASUREMENTS

SUBCONTRACTOR LIABILITIES (FROM 3.5.2)
Entry Type: Data Group

Composition:
SUBCONTRACTOR LIABILITIES + SUBCONTRACTOR LIABILITIES (FROM 3.5.2)

SUBCONTRACTOR LIABILITIES (TO 3.4)
Entry Type: Data Group

Composition:
SUBCONTRACTOR LIABILITIES + SUBCONTRACTOR LIABILITIES (TO

3.4)

SUBCONTRACTOR PERFORMANCE FILES
Entry Type: Data Element

SUBCONTRACTOR PROGRESS F.
Entry Type: Data Element / Local File

SUBCONTRACTOR RECORD FILES
Entry Type: Data Element / Local File

SUBCONTRACTOR SELECTION PROCEDURES
Entry Type: Process

This process includes: developing scope of sub-contract, comparison of pre-tender quotations, meetings with sub-contractors, final selection.

SUBCONTRACTORS PERFORMANCE FILES
Entry Type: Data Element

SUBMIT BILLS OF QUANTITIES
Entry Type: Process

Specification:

This process will deal with the following data: tender file, call for BoQs, tender details, staff requirement.

When the company tender is the lowest (or perhaps one of the lowest), it is usual for the bills of quantities to be called for analysis and checking by the client's consultants. The tender estimator drafts the bills of submission through the transfer of rates from the copy bills priced during the preparation of the tender. The chief estimator and directors meet with the estimator to discuss the policy to be adopted in pricing the bills, (eg, methods of incorporating profit, overheads and the fixed price allowance). They also examine the 'schedule of adjustments and agree the means by which these adjustments made at finalization should be incorporated into the bills.

The finalized bills are checked by the comptometer operators prior to submission and any arithmetical errors eradicated.

SUMMARIZE ESTIMATE
Entry Type: Process

Specification:

This process deals with the following data: priced preliminaries, build ups extensions & work sheets, measured

No.	Date	Superficial Cost
Site		S.M.
Start Date	Labourer	
Contract Period	Bricklayer	Carpenter

	Discounts	Totals
Preliminaries		
Labour		
Materials		
Plant		
Domestic Sub-Contractors		
Nominated Suppliers		
Nominated Sub-Contractors		
Fixed Price Allowance or Shortfall		
Totals		
Less Discounts		
Overheads		
Profit		
Insurance		
L.A. fees		
Dayworks		
Contingencies		
Provisional Sums		
Bond		
Add/Deduct tender adjustments		
Add/Deduct at Directors' discretion		
Tender amount	£	

Fig. D3. Tender summary and profit statement

rates, competitive late quotations, completed estimate.

A 'tender summary and profit statement' which is illustrated in Fig. D5, is part-completed prior to the tender finalization meeting. The tender number, date and site description are given, together with the proposed start date and contract period. The basic rates for labour, bricklayers and carpenters are also stated.

The 'totals' column is completed down to and including the line 'fixed price allowance or shortfall' and is then summed to give a gross total. The discounts on all subcontractors and nominated suppliers are listed likewise down the left hand column and again summed to give a total discount value. The total discount value is deducted from the gross total to leave a net cost figure on which profit and overheads will subsequently be based.

The overhead, profit and insurance lines are left to be completed at the tender finalization meeting, as is the 'bond' value. The value of the l.a. fees, dayworks, contingencies and provisional sums are filled in and will include for any associated markups.

Throughout the preparation of the estimate the estimator notes down conditions of the proposed contract that are unusual and onerous, together with risk items which have not been transferred to subcontractors or suppliers. These notes will be discussed with the directors prior to the finalization meeting should they pose serious problems, otherwise they will be raised at the finalization meeting for joint consideration.

The possible savings from competitive quotations received after the stated date of return are calculated. These savings will be brought up for discussion at the finalization meeting where they may be incorporated into the tender by a lump sum adjustment.

SUMMARY OF ORDER DETAILS
Entry Type: Data Element / Local File

SUNDRY COSTS
Entry Type: Data Element / Local File

SUPERVISION
Entry Type: Data Group

Composition:
SUPERVISION + FORMAL CONVERSATIONS + INFORMAL CONVERSATIONS + BONUSES + INVOICE VERIFICATION + PROGRESS TARGET

SUPPLEMENTARY REQUIREMENT TO EXISTING ORDER
Entry Type: Data Element

SUPPLIER DIRECTORY
Entry Type: Data Element / Local File

SUPPLIER ENQUIRY
Entry Type: Data Element

SUPPLIER ORDER FILES
Entry Type: Data Element / Local File

SUPPLIER QUOTATIONS
Entry Type: Data Group

Composition:
INTERIM APPLICATIONS + INVOICE QUERY + REVISED QUOTATIONS + SUPPLIER ENQUIRY + NEGOTIATION

These quotations are for materials listed either in the Bill of Quantities or from lists drawn up by the contractor. The quotations are analyzed and the prices absorbed into the contractors estimate. The quotations will usually offer a fixed price for a specified period.

These are used as the basis of orders & requisitions to suppliers at the construction stage.

SUPPLIER REGISTER
Entry Type: Data Element / Local File

SUPPLIER
Entry Type: Terminator

This terminator composes the following sub-terminators: competitive suppliers and selected supplier

SUPPORTING QUOTATIONS
Entry Type: Data Element

These are the competitive quotations under consideration.

SURVEYING DIRECTOR'S FILE
Entry Type: Data Element

SURVEYING DIRECTORS A.C. FILES
Entry Type: Data Element / Local File

Comment: A.C = Architect's Certificate

SURVEYING DIRECTORS CVC FILES
Entry Type: Data Element / Local File

SURVEYOR APPROVED COSTS
Entry Type: Data Element

SURVEYOR PROMPTS
Entry Type: Data Element

The contractor's surveyor sends prompts to ensure that all the necessary activities in process 3.2.3 are undertaken.

SURVEYOR'S CHECKS & CALCULATIONS
Entry Type: Data Element / Local File

SURVEYORS CVC
Entry Type: Data Element / Local File

SURVEYORS' FILES
Entry Type: Data Element / Local File

TAX EXEMPTION DETAILS
Entry Type: Data Element / Local File

TELEPHONE CALL
Entry Type: Data Element

TEMPORARY WORKS DETAILS
Entry Type: Data Element

TENDER COORDINATION MEETING
Entry Type: Process

Specification:

This process deals with the following data: quotations, preliminary schedule, unit rates & quotations, supporting quotations, established unit rates.

Prior to the final stages of the tender preparation the tender estimator and planner meet to discuss the tender and to bring the two sections of the estimate together. The meeting is informal and typically centres round the following points:

1 To run through the preliminaries schedule which is subsequently left with the estimator, together with supporting plant and scaffold quotations for pricing.

2 To exchange general impressions of the tender, its characteristics and problems.

3 To look at the pre-tender programme and the associated

method statements and discuss the implications of submitting alternatives to the client's completion date. (On occasions when an alternative completion date is to be submitted the planner usually produces a second programme and preliminaries schedule from which the time and cost implications can be measured).

The meeting serves to bring a measure of consistency to the tender preparations and gives both parties a fuller appreciation of the estimate.

TENDER DETAILS
Entry Type: Data Element

TENDER DOCUMENTS
Entry Type: Data Element

Tender documents will normally include specifications, tender drawings, Bill of Quantities, conditions of tender and proposed form of contract and any other relevant or unusual conditions.

TENDER ENQUIRY
Entry Type: Data Element / Local File

TENDER ENQUIRY ABSTRACTS
Entry Type: Data Element

The tender enquiry abstract lists page numbers in the bill of quantities, together with appropriate drawings, that should be sent with a particular group of enquiries, for example for all brickwork or plastering on a project. An illustration of this is given in Fig. D6.

TENDER ENQUIRY ABSTRACTS F.
Entry Type: Data Element / Local File

TENDER FILE
Entry Type: Data Element

All the data gathered and used during the tender process.

TENDER FILES
Entry Type: Data Element / Local File

TENDER FINALIZATION
Entry Type: Process

Specification:

The process deals with the following data: negotiated

savings & expected margins on S/C's, completed estimate, tender file, tender submission and pre-tender programme statement.

The process of finalizing the tender revolves around a two stage finalization meeting. Present at the first stage of the meeting are:

1 Managing Director (Chairman)
2 Surveying Director *
3 Chief Estimator
4 Tender Estimator
5 Tender Planner
6 Buyer *
7 Construction Manager

* Not always present for smaller tenders

The principal input to the meeting is as follows:

a. from the tender estimator:
   - Detailed knowledge of the tender
   - An analytical build up of the estimate
   - Provisional and P.C sum analysis sheet
   - Subcontract Comparison Sheet
   - Priced preliminaries schedule
   - Tender summary and profit statement
   - Any late competitive quotations

b. from the tender planner:
   - Detailed knowledge of the production aspects of the tender
   - Pre-tender programme
   - Method statement
   - Alternative programme and preliminaries schedule
   - Production related problems/restricted

c. from the buyer:
   - Probable buying margins on materials
   - Further negotiated reductions in subcontract tenders

d. from the construction manager:
   - Knowledge of the proposed methods and programme
   - Own impressions as to the minimum necessary content of the preliminaries schedule

When it is the intention to submit an alternative contract duration and price, the planner presents his alternative programme and preliminaries schedule for consideration and discussion.

Present at the second stage of the meeting are:

1 Chief Estimator
2 Tender Estimator
3 Managing Director
4 Surveying Director (major tenders only)
5 Buyer

On the basis of the agreed construction methods, programme, selection of subcontractors and adjustments established during the first stage of the meeting, the remaining personnel seek to finalize the mark-ups for overheads, profit and insurances and to agree the fixed price allowance if applicable. On large contracts the company chief executive will be consulted over

TENDER ENQUIRY ABSTRACT

Project .... S/S.C

BILL	PAGE	BILL	PAGE	BILL	PAGE	DRAWINGS
						SEND TO:

Fig. D1. Tender enquiry abstract

profit mark-up. Where similar types of work have been tendered for by other regions within the group, liaison with those regions might also take place with regard to the results of the tender, level of mark-up, etc.

TENDER INFORMATION
Entry Type: Data Element

TENDER SUBMISSION
Entry Type: Data Element

The tender is submitted to the client or his representative (usually a project manager or architect) by a specified date.

TENDER SUMMARY SHEET
Entry Type: Data Element

Summary of tender details including and observations or unusual features noted by the directors and chief estimator.

TENDER/VALUATION DETAILS
Entry Type: Data Element / Local File

TERMINATE HIRE NOTES
Entry Type: Data Element

TERMS OF PAYMENT
Entry Type: Data Element / Local File

TOTAL MATERIALS ON SITE
Entry Type: Data Element

TOTAL VALUE OF DAYWORKS
Entry Type: Data Element

TRADE FILES
Entry Type: Data Element / Local File

Comment: files on which reliable subcontractors are kept, by trade.

UNIT RATES & QUOTATIONS
Entry Type: Data Element

UPDATE CONTRACT PERFORMANCE RECORDS
Entry Type: Process

Specification:

This process deals with the following data: cost value comparison, previous CVC's file, CVC report (part).

Another aspect of a CVC report is a running summary of the actual value, overhead and profit from month to month against the forecast (target) values. The forecast values are taken from the contract turnover forecast sheets which are normally completed in the initial weeks of the contract and the actual values are taken from the CVC just completed in section 3.4.3.4 in DFD No 3.4.3.

VALUATION DATES
Entry Type: Data Element

VALUATION FILES
Entry Type: Data Element / Local File

VALUATION MEETING/ SUBMISSION
Entry Type: Process

Specification:

This process deals with the following data: complete valuation, analyzed valuations, final value to be certified, notify certifying value & valuation date, notification of value to be cert & val date.

The dates of the valuation meetings are normally agreed with the pqs at the beginning of the contract and may occur at any time during the month. This means the surveyors who may be responsible for more than one contract can spread their workload evenly over any month. The valuations are usually presented by the more senior surveyors with the presentation taking the form of a day-long meeting with the pqs.

It is normal practice for the company surveyor to accompany the pqs on a walk round the site prior to the valuation. This enables the surveyor to point out work that may have occurred for the first time on the contract, and also items of daywork and variations which may be a source of contention during the valuation. The primary reason for the walk round site is to update the pqs with the progress on the contract which he may not have visited since the last valuation.

The valuation itself normally involves the surveyor and pqs running through the pre-prepared company valuation. The pqs challenges items he considers over measured or not in line with his own interpretation of the bills of quantities. Any changes are usually written into the valuation during the meeting although section totals and the summary may be left for correction until the next day. Disputed items may be noted for verification on site later in the day or may be left for the pqs to consider on his own perhaps with reference to the clerk of works or architect.

The PC sum items for nominated subcontractors and suppliers are usually reviewed with the pqs finalizing his decision as to the amount each should be certified. (The pqs should have normally received copies of the relevant interim applications and invoices prior to the valuation enabling him to verify them with the appropriate consultants).

The variation and daywork sections of the valuation usually contain the majority of disputed items. Some dayworks may not have completed the authorization and pricing process and the certification of monies against these items will be left to the discretion of the pqs. The valuing of variation items may need checking by the pqs who may not be prepared to include the full value claimed into the valuation. The company surveyor will attempt to provide sufficient back up information such as rate build ups, quotations, etc, to support the major items and aim to get the full value certified as soon as possible.

The extent of disputed or changed items and the normal practice of the pqs usually determines whether the valuation is finalized and agreed during the meeting. If not, the pqs normally telephones the final details to the company surveyor the following day.

Invoices are not submitted by a contractor but the pqs copy of the valuation and the summary sheet acts as a statement.

The contractor's surveyor is normally required to indicate to the surveying director when a valuation has taken place and to state the expected cheque value. This is typically an informal procedure done by word of mouth during the day after the valuation.

VALUATION SUMMARY
Entry Type: Process

Specification:

This process deals with the following data: preliminaries totals, complete valuation, fluctuation totals, total materials on site, total value of dayworks, variations total, work section (trade) total, section totals.

As each section of the valuation is prepared and finalized, normally section totals are calculated to carry forward into a single page summary sheet. Each section is itemized and totals listed, the sum of them being the gross application for payment. The summary sheet acts as a statement to the client but may need re-drafting should the application be altered significantly due to disputed items.

VALUATIONS FILES
Entry Type: Data Element / Local File

VALUE TO BE CERTIFIED
Entry Type: Data Element

VARIATION FILES
Entry Type: Data Element / Local File

VARIATION ORDER
Entry Type: Data Element

VARIATIONS TOTAL
Entry Type: Data Element

VARIATIONS/ INSTRUCTIONS F.
Entry Type: Data Element / Local File

VERBAL INSTRUCTIONS
Entry Type: Data Element

Usually issued on site

VERIFIED BUILD UPS & EXTENSIONS
Entry Type: Data Element

VERIFIED INVOICE
Entry Type: Data Element

VERIFY INTERIM APPLICATIONS
Entry Type: Process

Specification:

This process deals with the following data: dayworks, internal assessment of s/c progress, subcontractor record files, variation files, valuation files, checked application, priced dayworks, interim applications.

Upon receipt of a direct subcontractor's interim application the surveyor will check the details and measurements. Payments are not made 'on account' with each application having to be verified. If verifiable the application is passed for payment. If not, it will typically be altered by the surveyor and the subsequent payments made will be based on his own records.

The value of works under the category of variations may be assessed and paid 'on account', although every endeavour is normally made to agree such items as soon as possible. When possible subcontractors are usually requested to submit quotations for work variations which if acceptable form the basis on which interim payments are made.

Dayworks are normally checked against the general foreman's records and the surveyor seeks to ensure that the details on the sheets whether signed or not give a true reflection of the hours and resources used. The pricing of dayworks will be checked against the rates previously agreed and included in the subcontract order. Verified dayworks will be accepted and included in the next payment. Dayworks which cannot be agreed or substantiated may be totally rejected or reduced to a realistic value.

note: Direct Subcontractor = Specialist Trade Subcontractor = Specialist Trade Contractor = Contractor's Subcontractor.

VERIFY INVOICE
Entry Type: Process

Specification:

This process deals with the following data: price books, invoice & posting slip, invoice verification, invoice query, verified invoice, invoice payment/ rejection, copy material returns, copy orders.

Invoices are normally checked firstly against the material orders and secondly against the relevant material return sheet. An invoice is checked for rate against the order and delivery and quantity against the return sheet. An accounts department will also have copy 'call off' sheets which they can use to check a delivery further if necessary. Invoices which relate to materials not specifically priced on the covering order (e.g. light sundry items from the local merchant) are verified using a price book.

Invoices not fully verifiable are referred to either a buying department (if a covering order cannot be found) or site (if the materials in question are not recorded as having been delivered). Those which still remain unverifiable are usually rejected and the supplier is notified by letter and requested to submit a credit note and a revised invoice.

VISIT REPORT
Entry Type: Data Element

The purpose of this visit is to assess the calibre of the supplier's managerial staff and systems, the quality of his production equipment, his ability to meet peak demands, his storage and delivery service, and above all his quality control & general reliability.

VISIT SITE
Entry Type: Process

Specification:

This process deals with the following data: site inspection report, site visit report, site visit notes.

As soon as it is practical after the award of the contract the construction manager and agent visit the site. It is useful to have a standard site inspection report form which lists the information looked for during this visit (although the form may not always be completed). The principal headings under which information is usually gathered are:

1 Site features
2 Services availability
3 Labour appraisal
4 Ground conditions
5 Details of local suppliers/subcontractors
6 General comments

WEEKLY PROGRAMME
Entry Type: Data Group

Composition:
WEEKLY PROGRAMME + RESOURCE FORECAST

WEEKLY PROGRAMMES
Entry Type: Data Element

WEEKLY PROGRAMMES /AGREED ACTION
Entry Type: Data Group

Composition:
WEEKLY PROGRAMMES + AGREED ACTION

WORK MEASUREMENT
Entry Type: Process

Specification:

This process deals with the following data: subcontract orders, subcontractor liabilities, work record & measurement. progress measurements, progress measurement F., dayworks, internal assessment of s/c progress, subcontractor record files, daywork records, allocation sheet.

Throughout the execution of each subcontract a surveyor usually records progress through detailed measurement of the works. The measurements (obtained from physical site measurement) are usually used in the preparation of the monthly valuation for the architect certificate but may also occur throughout the month possibly weekly, depending upon the payment terms of the subcontract orders.

The measurements are normally used together with the daywork records received from the agent to calculate payments due to subcontractors and to calculate the monthly liability for the main account costs.

Many of the larger subcontractors used by a company may employ their own surveyor who periodically arranges to meet the contractor's surveyor to agree both interim and final measurements and the value of interim applications.

The contractor's surveyor receives daily allocation sheets from the site agent which record the work undertaken by direct labourers.

WORK MEASUREMENTS
Entry Type: Data Group

Composition:
WORK MEASUREMENTS + MEASUREMENT OF VARIATIONS

WORK RECORD & MEASUREMENT
Entry Type: Data Element

WORK SECTION (TRADE) TOTAL
Entry Type: Data Element

WORK SUPERVISION
Entry Type: Process

Specification:

This process deals with the following data: daily diary, 8 week programmes, contract programme F., subcontractor progress F., daywork records, allocation sheet, weekly programme, agreed course of action.

The direct supervision of the work on site is usually the responsibility of the general foreman and site agent. The site agent usually is more involved with the overall coordination of all the resources and hence spends less time out of the office. The general foreman is responsible for directing and motivating the workforce and aims to ensure the agreed production targets are met.

Subcontractor progress meetings are most often held monthly with the nominated subcontractors and principal direct subcontractors. The meetings with nominated subcontractors are usually held on a formal basis prior to the monthly main contract progress meeting. A 'standard agenda' is best followed in the meeting which seeks to gather information that can be fed into the main progress meeting. The appropriate consultant from the design team usually attends such meetings which are minuted by the construction manager. The meetings with principal direct subcontractors (contractor appointed) are generally less frequent and are not always minuted in the same formal way. Such meetings normally deal with technical rather than financial matters.

The general foreman is usually required to record the day to day labour strengths of all the subcontractors on site onto a daily diary sheet. A more formal record may be kept and given to the resident engineer/clerk of works if requested.

Dayworks undertaken by the subcontractor are signed for and agreed by either the agent or general foreman. The details of such daywork is usually passed to the contractor's surveyor, together with a copy of the appropriate daywork sheets for payment purposes. The site agent is responsible for completing the company daywork sheets which in due course would be presented to the architect (see section 3.2.6 in DFD No 3.2).

Direct labourers and tradesmen have their hours and work activities recorded by the general foreman onto daily allocation sheets. These sheets are used to record the hours worked by each man and to allocate the hours to specific tasks. The allocation sheets are passed to the surveyor, the company wages office and bonus surveyor, as a record of work done.

Any variation to the contract works which affect the works of subcontractors are conveyed to them on an instructions to subcontractors form. These forms are filled out by either the surveyor or the construction manager and forwarded to the subcontractor as soon as possible after the issue of the instruction.

WORKS/YARD VISIT NOTES
Entry Type: Data Element

WRITE UP & DISTRIBUTE MINUTES
Entry Type: Process

Specification:

This process deals with the following data: minutes & action notes, minutes agenda & action notes, job review meeting files, manager's minutes, buyer's report.

The construction manager is usually required to circulate the minutes of the meeting to the directors and to all those who were present.

The buyer also produces brief minutes of the sections concerning materials and possibly subcontractors. These minutes are also circulated to the directors and the parties present at the meeting.

'ONE OFF' PAYMENT AUTHORIZATION
Entry Type: Data Element

8 WEEK & WEEKLY PROGRAMME F.
Entry Type: Data Element / Local File

8 WEEK & WEEKLY PROGRAMMES
Entry Type: Data Element

```
8 WEEK PROGRAMME
Entry Type: Data Element / Local File


8 WEEK PROGRAMMES
Entry Type: Data Element / Local File
```

4. Developing more robust systems

Introduction

The first three chapters of this book have described a method of analysing and improving the flow of management information related directly to the activity of construction in a contracting company. The SDA technique has been used to follow and map the route of individual pieces of data as they have flowed round a company's management system. The DFDs show how the data move around a system and are transformed as they do so. It shows the source of the data, their structure, how they were distributed and their final destination. The DD provides a comprehensive description or specification of all the data. It is argued that the careful monitoring of the efficienty of data systems is important if managerial efficiency is to be increased (ref. 24).

Construction project managers, according to Karlen, often spend more than 70% of their time dealing in one way or another with data. So any improvement in the system of management information a manager uses will directly affect his efficiency. Furthermore, such an approach will offer the 'system benefits' of consistency and reliability which will lead to a reduction in risk, as has been demonstrated in a number of the high risk industries behind which construction lags in terms of system development. This chapter looks at the benefits, possible drawbacks and the future of SDA.

Possibilities and benefits

A number of large international contractors (see chapter 1) are beginning to see the benefits of computer-integrated construction project management and in particular site management. Such advances are becoming less confined to the major companies, which have usually had to finance considerable system development costs. Many of these often very expensive systems have been designed and constructed using an approach that in effect begins with a logically ideal solution and starts its design from a clean sheet of paper. In practice this solution needs much development and amendment to accommodate any unforeseen operational constraints.

The SDA approach as outlined in this book will allow the construction project manager much more control over the development of the system he will eventually have to manage. This is achieved in two ways: by the development of an efficient and effective manual or part computerized system (as advocated in chapters 1 and 2); and by computerizing, in a controlled and incremental way, units of the system developed. This means that the often large development costs can be kept down. New system units are simply bolted on at the interface points. With cheaper, more powerful and more reliable hardware, and the more effective system and design technique outlined in this book, the concept of computer-integrated systems now comes within the realms of possibility for a much larger range of contractors.

Further development of the concept – a better system

Whether or not the step is taken to computerize, considerable benefits can be gained from better and more efficient management information systems. Aviation and aircraft technology provides an analogy. After the First World War, aviation moved from seat of the pants flying to seat of the pants flying with 'flying aids', such as navigational aids, bomb-aiming techniques and improved instrumentation. After the Second World War, as aircraft and aircraft safety became more complex, manuals of procedures and techniques became common- place. Computing capability has brought an advance in two areas

(a) information management systems which have gradually become more sophisticated through developments such as the autopilot, to the extent that today's aeroplanes can apparently take off, fly and land without any human pilot intervention
(b) computer-driven simulators, run by systems which can emulate flying knowledge, flying skills and the effects of the elements; these simulators are now so effective that experienced pilots find them 'frighteningly realistic'.

The UK construction industry could be said to be broadly at the point where the first commercial jet airliner, the Comet 4b was introduced: seat of the pants intuitive management but with some aids, such as procedures and systems.

This book stresses that better systems will bring productivity benefits, reliability and consistency, and will free the manager to produce excellence, as he becomes more confident about the data he deals with, seeks, or supplies and on which he makes decisions.

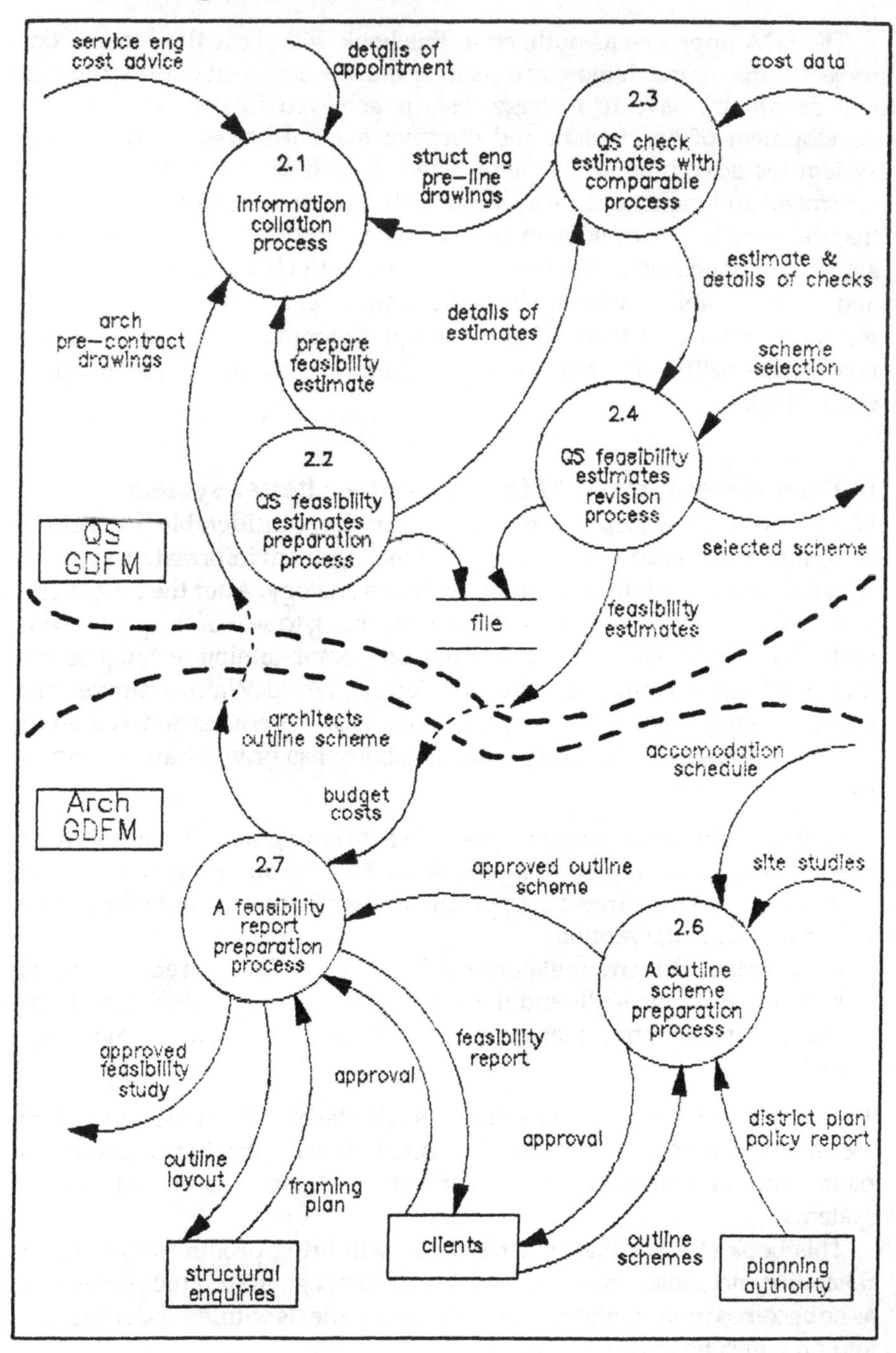
service eng cost advice
details of appointment
cost data
2.3
QS check estimates with comparable process
struct eng pre-line drawings
2.1
Information collation process
estimate & details of checks
arch pre-contract drawings
details of estimates
prepare feasibility estimate
scheme selection
2.4
QS feasibility estimates revision process
2.2
QS feasibility estimates preparation process
QS GDFM
selected scheme
file
feasibility estimates
architects outline scheme
accomodation schedule
Arch GDFM
budget costs
2.7
A feasibility report preparation process
approved outline scheme
site studies
2.6
A outline scheme preparation process
feasibility report
approved feasibility study
approval
district plan policy report
approval
outline layout
framing plan
clients
outline schemes
structural enquiries
planning authority

Fig. 1

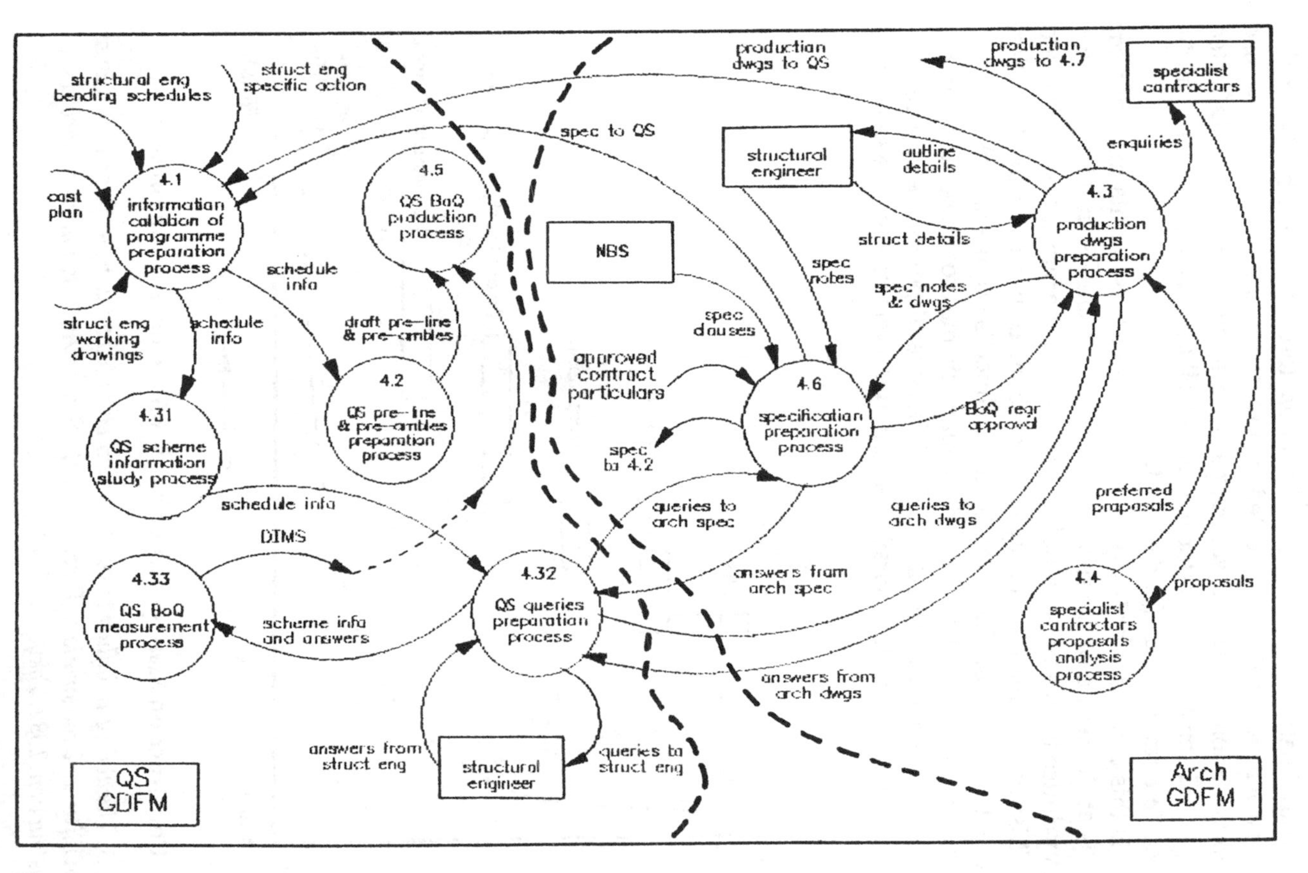

structural eng bending schedules
struct eng specific action
cost plan
4.1 information collation of programme preparation process
struct eng working drawings
schedule info
schedule info
4.31 QS scheme information study process
4.5 QS BoQ production process
draft pre-line & pre-ambles
4.2 QS pre-line & pre-ambles preparation process
schedule info
DIMS
4.33 QS BoQ measurement process
scheme info and answers
4.32 QS queries preparation process
answers from struct eng
structural engineer
queries to struct eng
QS GDFM
spec to QS
production dwgs to QS
production dwgs to 4.7
NBS
spec clauses
approved contract particulars
spec to 4.2
4.6 specification preparation process
queries to arch spec
answers from arch spec
queries to arch dwgs
answers from arch dwgs
structural engineer
outline details
struct details
spec notes
spec notes & dwgs
BoQ reqr approval
4.3 production dwgs preparation process
specialist contractors
enquiries
preferred proposals
proposals
4.4 specialist contractors proposals analysis process
Arch GDFM

Fig. 2

Management of data system interfaces

The *2001 series* (ref. 6) identified a number of clients' concerns about the construction industry and the way it performs. Foremost of these was the adversarial nature of the industry. This was closely followed by the level of concern felt about the fragmented nature of the industry and how bad communciation between the various parties to a construction project affected design and construction performance, as well as the overall quality of the project in terms of value for the client.

It is this data transfer between contractual parties and its management that is at the centre of the second concern. For example, in Fig.1, the two heavy dotted lines represent the boundaries of, or the interface between, two GDFMs: that of a quantity surveying practice and of an architectural practice. Two data flows cross the interface: the architect's outline scheme, and the feasibility estimates. The architect's outline scheme crosses between proces 2.7 (architect GDFM) and 2.1 (QS GDFM). Several points are apparent on examining this data flow. Firstly that the two systems give the same data

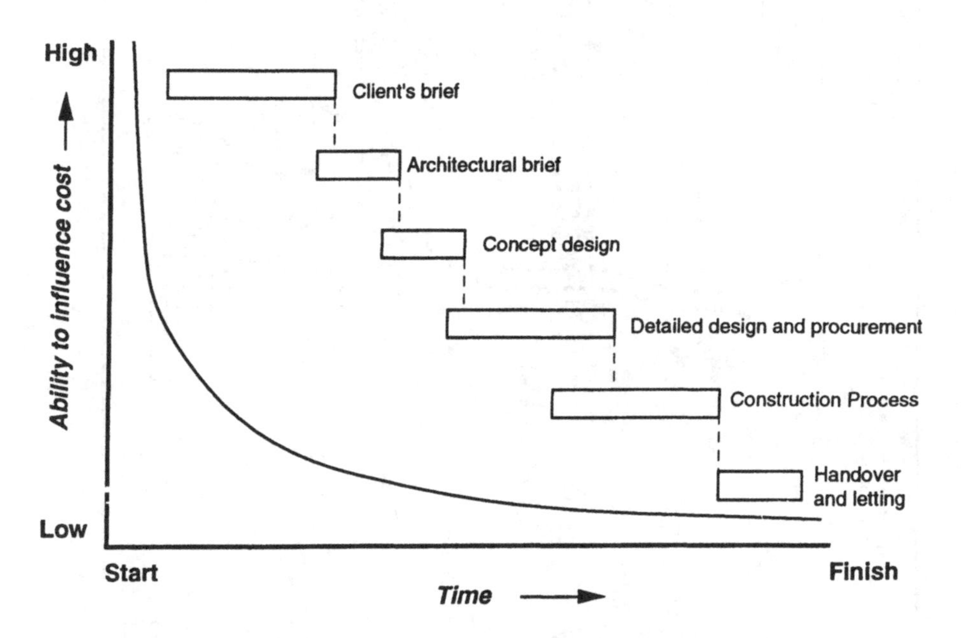

Construction Management phases for a construction project

Fig. 3. The ability to influence construction costs (adapted from Tucker, R. Const. motability and automation. Proc. 9th Int. Symp. Automation and Robotics in Construction, 1992, Tokyo, pp. 1–7)

different names: architect's outline scheme by the architect GDFM; and architect's pre-contract drawings by the QS GDFM. Secondly, the names imply a different perception of what is being transmitted. To the architect it is the concept that is being transmitted: creation, a new building, the drawings being only one way to express ideas. Others might be a three-dimensional model, a sketch or a textual description. To the QS the data are clearly just 'drawings' and a 'specification', possibly the new material a QS needs to prepare his main contribution, namely: the Bill of Quantities.

Feasibility estimates cross between processes 2.4 (QS GDFM) and 2.7 (architect GDFM). Again, this flow of data has different names, each side of the interface reflecting different cultures, value systems and aspirations. It is called feasibility estimate by the QS GDFM and budget costs by the architect GDFM. Fig. 2 shows six interface points between two different, but similar architect and QS GDFMs, and there are greater similarities between the two GDFMs when naming data flows.

An analysis of this type together with the supporting data available from the DD provides a clear picture of the nature, structure and content of the data that are transmitted across an interface. It allows not only all the interface points to be clearly identified and recorded but also the nature of the data being transmitted, and key issues surrounding the sending and receiving, to be understood. The processes provide the construction project manager with the information necessary for the better management of data system interfaces.

A systems approach to data management – the key to future development in IT

IT at its most basic is about the electronic processing, manipulation and transmission of data. If the industrial revolution was chiefly about extending the human muscle and arm, then the electronic revolution is about extending the human brain - extending it in terms of the ability to assimilate and process data, to store and make decisions based on stored knowledge and experience, and to transmit data quickly and efficiently. It is the speed, capacity and increasingly the reliability of the new IT devices that will make the biggest impact on the way we currently operate. Advanced IT solutions will be more than just cost effective; they will, when well harnessed, offer considerable competitive advantage to the 'first company in', with other companies being forced to follow if they are to remain competitive. The advantage will lie in the ability to offer more flexibility and achieve considerable productivity gains.

The ability to influence construction costs is demonstrated in Fig. 3. Once some of the techniques currently used in other industries are properly deployed in construction, construction costs will fall significantly. This is particularly so at the client's brief stage (identifying a business opportunity, developing the business proposition which offers best value for money, designing the business proposition as a system achieves optimum performance standards acquired by experience), and at the architect's brief and concept design stages.

The new-found ability will optimize a concept design and components at a point which will affect construction costs dramatically. It will be at precisely this point, for example, that major Japanese and other manufacturers will be able to (in fact already can) relate to construction (see chapter 1) and use their flexible manufacturing and simultaneous engineering techniques to improve value for money through automation.

Such reductions in construction costs, once they are demonstrably possible, will be demanded by the major sophisticated, regular clients of the industry. This will also affect the way buildings are designed and built, or perhaps fabricated. All major and most middle-range contractors will have little alternative but to adapt to different ways of working. A systemized and componentized building will have to be built in a very different way. Similarly the management of the design *and the construction* will be very different.

This book has outlined a useful and practical first step in systemizing the management of construction by a contractor. If a contractor considers the construction of major new projects to be a core business activity, then he has little alternative but to take these developments seriously immediately, as to develop such a capability will take time. Perhaps more important is the need to develop staff skills and, probably more difficult, the mental/intellectual approach to cope with these techniques. It will take time, effort and, of course, money. It will also require vision and courage on the part of senior managers and directors of the major contracting companies.

Principal benefits of the structures data analysis technique

This book has developed, explained and argued a number of themes. Perhaps the most fundamental is the need to develop 'systems thinking', in particular as it relates to IT. Over the past nine years, more than 150 experienced senior construction project managers taking the University of Reading Master's Programme in project management, have been persuaded, sometimes reluctantly, to view the world around them, and in particular their company, project and job, as a hierarchy of systems. Almost without exception they have done so, and achieved a clearer understanding of often very complex business and management situations as they relate to data and issues of communication, good or bad.

The holistic, top-down approach of SDA and its ability to partition in a rule-driven way complex 'real world' management information systems into 'manageable bits', is another benefit, and minimizes investigator influence on results. A third benefit lies in what the system records and the manner in which it is recorded. It allows systems comparison, systems improvement and the examination of quite specific but key or critical areas such as data interfaces. Another important benefit stems from the rule-driven methodology. Because of the partitioning, numbering and naming systems used, it is possible for different investigators to tackle different aspects of the same system, and to be able to relate easily one sub-system to another. These features ensure that the work produced by the SDA is broadly repeatable and thus refutable, and that for a properly trained and experienced investigator, bias is reduced to a minimum.

Shortcomings of the technique and the approach

The technique is a rational way of looking at aspects of an apparently irrational world. While Cray super-computers are finding clear patterns from masses of data on the world's weather and developing the concept of the planet's weather system, for the intuitive, spontaneous, creative, unique human being, this rational approach is not always easy to accept. The technique is labour-intensive and slow, even using a computer-driven 'systems analyst workbench'. It does require a mental jump although neither the technique nor the concept is particularly difficult intellectually. It offers just a snapshot in time, and takes no account of the evolving nature of a system. Similarly, it shows only that data flow down a particular route, and not the rate of flow or the sensitivity of the data that are flowing.

A final short-coming is that SDA takes a very narrow view. It takes no account of the human dimension, indubitably the most important. Neither does it take any account of time, influence, responsibility or authority. Results must therefore be viewed in the light of these shortcoming, as must the fact that the results, and the GDFM described in this book, represent a relatively early stage in on-going research. As more DFMs are looked at and compared with the GDFM, and as more GDFMs are combined or bolted together to develop a structured data system map (SDSM) of the industry, results will in time become more robust, and therefore more reliable.

In conclusion – towards 2001

The next eight to ten years will probably see greater advances than, for example, the emergency of the micro-computer in the late 1970s. Change or progress is driven by breakthroughs in knowledge or better devices, usually for some kind of measurement or techniques (ref. 37). Communications will almost certainly get faster and cheaper as a result of fibre optics and the availability of bigger cheaper microprocessors. The bigger cheaper microprocessors will also encourage more sophisticated operating systems where computers will think more like humans, and be able to learn from experience. Bigger and more sophisticated data bases will store and use 'learned knowledge' in a way that begins to replicate methods used by an experienced architect or civil engineer, for example. This approach has already been used in the automotive industry with some success. It is known as knowledge-based engineering (KBE). KBE ensures that a computer-driven design system will not start as if from a clean sheet of paper when developing a design, but will work with a human designer as a team, using lessons learned from its 'knowledge' of previous projects. For example, if an architect is designing a hotel, the computer-driven design system would begin by designing a 'processing system', with basic data, such as size, capacity, standard of service, quality of hotel, supplied by the architect.

An airport terminal is a good example of a building where the client's brief could be designed in this way. At its most functional an airport terminal is a 'people-processing system'. People enter randomly, pass through a series of filters (customs, security, immigration), are sorted by destination and departure time, and then dispatched in secure batches. The reverse is true 'air-side', as people proceed through the system and filters to reach 'land-side', where they depart randomly. There are also sub-systems within the processing system, people who are part of the sorting, filtering and dispatching system, who enter, work in and leave the airport terminal system. KBE will develop, in partnership with a designer, a data-driven model which can also be shown graphically, and will be governed by performance criteria and the KBE system ability to achieve optimum solutions.

The KBE design will form part of the more comprehensive and realistic brief offered to a concept architect and other designers. In effect it gives 'reserved space' around which designers can design something architecturally acceptable, with structural integrity as well. This approach is close to new practices in the car industry, where skin and interior designers are given reserved space which they have to accommodate in their designs. Airport terminals and hotels, of course, have people working in and passing through them. So there are important social and human factors to be accommodated. It will be possible, however, for much of this to be entered into the 'knowledge/experience' of a KBE system. KBE will have a second contribution to make at the stage when the client's brief is turned into an architectural brief, at the concept stage and at the stage where the architectural engineering or working details are produced. It is likely that the KBE system will also be able to design components (or use standard components) and optimize them to drive out hidden quantity. Once these components are designed and put together into a building (system) or a computer-driven 'prototype' that can be tested and found to work (both in a holistic way and at each key point), it will be possible for the system to control an automated production line for the components.

All this may seem a long way off, but chapter 1 gives examples of where this approach in its entirety is already being tried. What is certain is that major contractors will be forced by the market to change dramatically the way in which they manage and construct buildings, and it will happen well within the working lives of those currently in the industry. If this challenge is not fully and wholeheartedly met, the future for today's major contractors as serious international construction companies will be in some doubt. It does well to remember the fate of the Swiss watch industry and the British motorbike industry. It is also worth pondering the high levels of productivity achieved in the UK by Japanese car companies such as Nissan. The management information process will force managers and directors to come to terms with the concept and the necessary changes. Certainly the Japanese car companies have been very successful at perfecting the paper-based system before they began to computerize it. If techniques such as KBE are used properly in construction, the clients of the industry will be the first to benefit from the road to automation, in lower construction costs and better value for money. But it will not be only the client who benefits from what should be given the generic title of computer-integrated construction. Lower construction costs will make a significant number of unattractive or marginal projects a viable, attractive business proposition. This will lead gradually to a bigger workload for the industry, from which every part of the industry will benefit. There is also just a chance that such high-tech developments will improve the image of the industry and will help it to attract its rightful share of the most able young people. This is, of course, crucial for the long-term well-being of the industry. Any improvement in image brought about by a changed reality will address the problems that often cause new entrants to the workforce to consider all other industries before construction. The Japanese should have the last word, as they succinctly sum up the recruitment problems of the industry with five 'K' words: *Kitanai* (dirty); *Kitsui* (hard); *Kiken* (dangerous); *Kiyujitsu ga sukinai* (too few holidays); *Kyukyo ga yasui* (not enough money).

Appendix 1. The structured data analysis technique

Introduction

Chapters 3 and 4 describe a general data flow model (GDFM) constructed from data flow models (DFMs) of three large contracting companies. The GDFM consists of two principal parts: the data flow diagrams (DFDs) and the data dictionary (DD). To understand the fieldwork methods, the model itself and the benefits of looking at construction activity in this way, it is necessary to understand first a little about the technique used, and the way in which the resulting model is presented. Appendix 1 therefore explains the structured data analysis (SDA) technique, describes what SDA records, and outlines its benefits and limitations.

Appendix 1 is therefore necessarily theoretical, as it also seeks to establish a place for the SDA technique within the relevant existing body of knowledge as currently available in the 'public domain' literature. This research is both necessary and important, as it provides the foundation upon which the study, and its model and conclusions are built. As was mentioned in chapter 1, current scientific opinion (ref. 37) suggests that to develop 'good science' there needs to be a balance or harmony between good theory that leads to the development of 'ideal' models, and good practice, experimental empirical work that draws principles from an analysis of the 'best of current practice' and allows a debate with the theoretician. As was stated in chapter 1, the 'tension' that at times exists between the two approaches enables scientific opinion to suggest that new understanding can and usually does occur.

SDA is a rule-driven research tool developed originally by IBM for systems analysis work in the electronic data processing (EDP) industry. The technique enables the recording and mapping of flows of data within a defined system and between 'processes': i.e. points within that system at which data is altered in some way. Suggestions are made later in this Appendix as to where the SDA technique fits in the broader systems movement. It is argued that SDA is a 'hard system' as defined by the current 'systems paradigm', and that it has certain advantages over other techniques. Modifications to the EDP version of the technique needed when it is used for management systems research are outlined. In conclusion the possible benefits and limitations of the technique are discussed.

SDA provides a clear and consistent rule-driven technique that allows management system researchers to describe data flows within organizations, and as a result to compare important aspects of organizations. Such comparisons provide a new and potentially beneficial avenue of research.

The flow of data in construction organizations

A number of studies have demonstrated that the flow of data between the parties to a construction project is a major component of construction management activity. The work of Munday (ref. 40) suggests that managers involved in the activity of construction spend nearly half their working time on tasks devoted exclusively to the transmission of information (data). He also found that most of a construction manager's work was made up of information processing and management. From the results of his analysis, Munday suggests that the flow of data appears to be particularly critical in the construction industry, given the unique nature of each project and the fragmented nature of the industry itself (ref. 41).

This view is confirmed by, for example, Karlen (ref. 24). Ball, Bowley and Bishop (refs 42-44), have also suggested that the construction industry is unduly fragmented and that this has hindered progress and innovation, and has adversely affected the way the industry (and its clients) views itself and the service it should provide. Bishop attempted to counter this industry fragmentation with a 'systems map' that both showed how the industry worked and illustrated the divisions that existed. To date, Bishop's systems map (or description) remains largely unchallenged or built on, possibly because a suitable tool has not been available to make the 'surge in development' of a particular avenue of research that Porter *et al.* (ref. 45) and Kuhn (ref. 24) describe typically happens when a new investigation or measurement tool is developed.

The systems movement

In his excellent book that describes the shape of the systems movement (ref. 46), Checkland (1981) identifies just one area: 'problem solving development'. This is a sub-set of systems thinking, an approach that allows the recording and mapping of data. He calls this hard systems, and argues that it has been influenced strongly by work in engineering, control theory and information theory. Checkland identifies two divisions in the hard systems sub-set: systems engineering as typified by chemical engineering and pro-

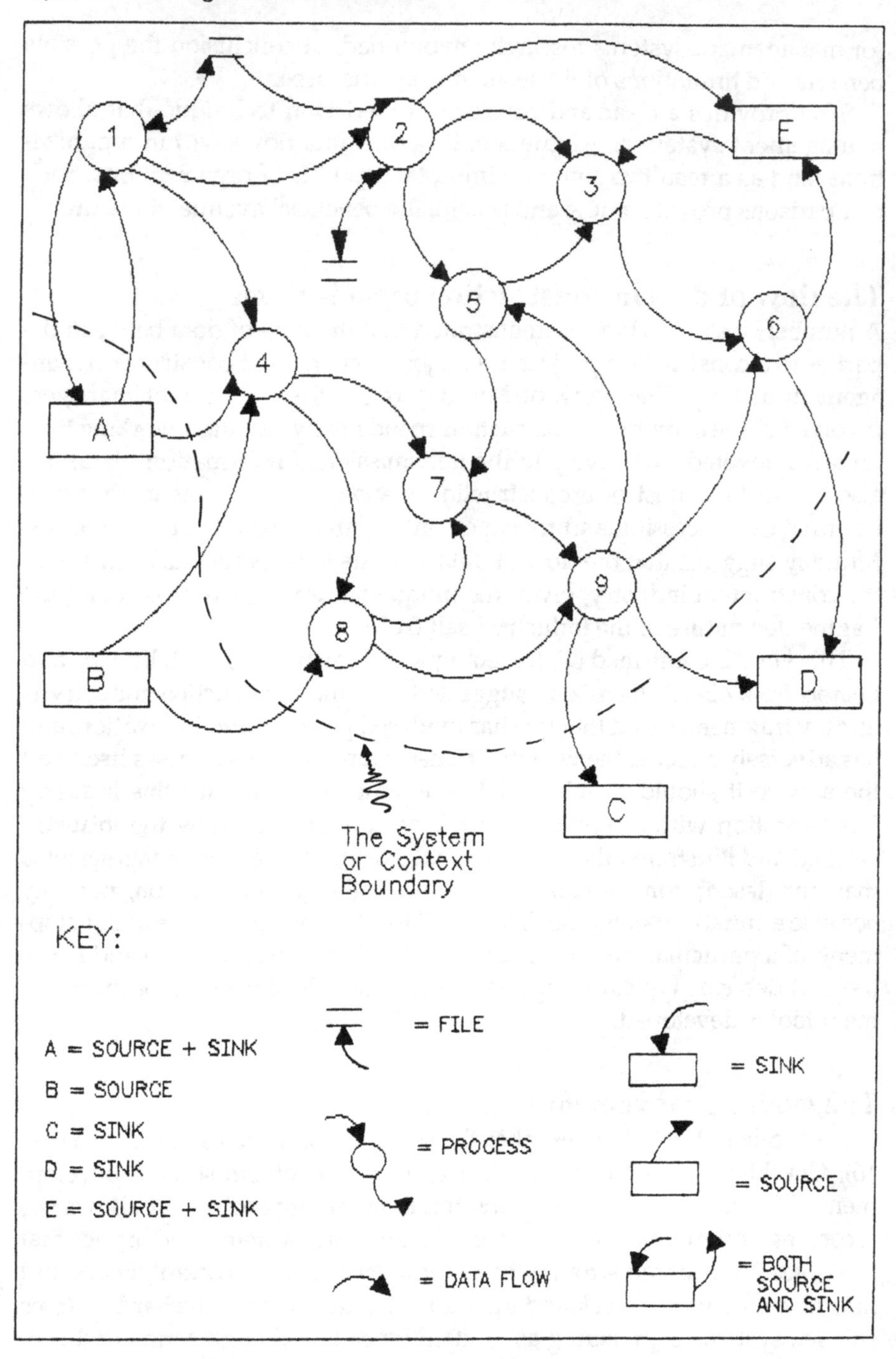
1
2
3
4
5
6
7
8
9
A
B
C
D
E
The System
or Context
Boundary
KEY:
A = SOURCE + SINK
B = SOURCE
C = SINK
D = SINK
E = SOURCE + SINK
= FILE
= PROCESS
= DATA FLOW
= SINK
= SOURCE
= BOTH
SOURCE
AND SINK

Fig. A1

cess engineering; and computer systems engineering and systems analysis. This Appendix will consider computer systems analysis.

Checkland argues that a computer systems analyst attempts to use the methodology to build a model of an observed phenomenon, and then build a logical model of a future improved system from it. This view is supported by systems analysts such as Jackson (ref. 47) and De Marco (ref. 48). A survey of the appropriate literature (refs 47-52) suggests that there are several key works on systems analysis, which fall into two distinct groups: those that base their analysis on 'data structure' (refs 47, 49, 50), and those that base their analysis on 'data flow' (refs 48, 51).

As mentioned in the introduction, this Appendix is concerned with tools that map or measure the transmission or flow of data on construction projects between the various parties (but in a holistic way), and that record but eliminate or minimize any of the effects of the industry's fragmentation. The technique that bases its analysis on data flow is more appropriate than the data structure approach, which as Jackson admits builds an analogue model that behaves outwardly in the same way as the observed data system. By contrast the SDA approach maps flows of data, their passage through an observed system, their transformation and co-ordination, exactly as they are observed, and in the process records a much more complete picture. SDA is interesting because it focuses on information decisions, which are important for practising managers. Its use in construction management systems research was first outlined in Fisher (refs 12, 26, 53-57).

SDA, as first developed for application in EDP, consists of a number of related analysis tools. The first is a technique to assist in the partitioning of the overall system and to document that partitioning clearly. For this purpose the DFD has been developed. This is, in effect, a network of interrelated processes expressed graphically. Fig. A1 shows a typical DFD. It follows a convention of showing flow of data but not control. Thus it differs from the flow chart, which portrays data from the point of view of those who act on those data. The DFD records the observed flows from the point of view of the data themselves.

Four notation symbols are used in a DFD

(a) the named vector (data flow) which portrays a data path
(b) the bubble: a circle representing a process which portrays transformation of data
(c) the straight line which portrays a file, or data base
(d) the box: a data source or sink, which portrays a net originator or receiver of data, typically a person or an organization outside the subject system context (the domain of a particular study).

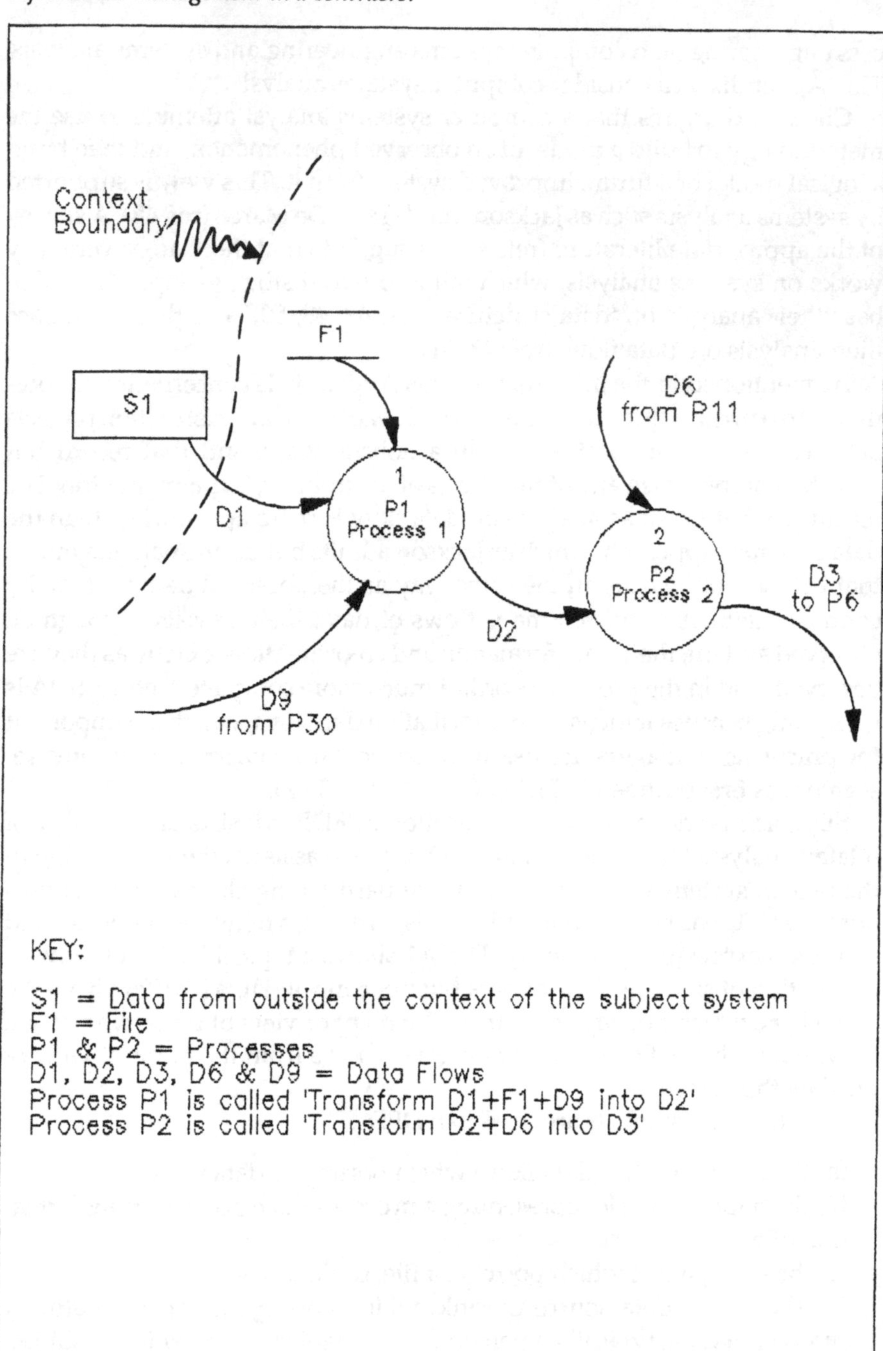
Context Boundary
F1
S1
D6 from P11
1 P1 Process 1
D1
2 P2 Process 2
D3 to P6
D2
D9 from P30
KEY:
S1 = Data from outside the context of the subject system
F1 = File
P1 & P2 = Processes
D1, D2, D3, D6 & D9 = Data Flows
Process P1 is called 'Transform D1+F1+D9 into D2'
Process P2 is called 'Transform D2+D6 into D3'

Fig. A2

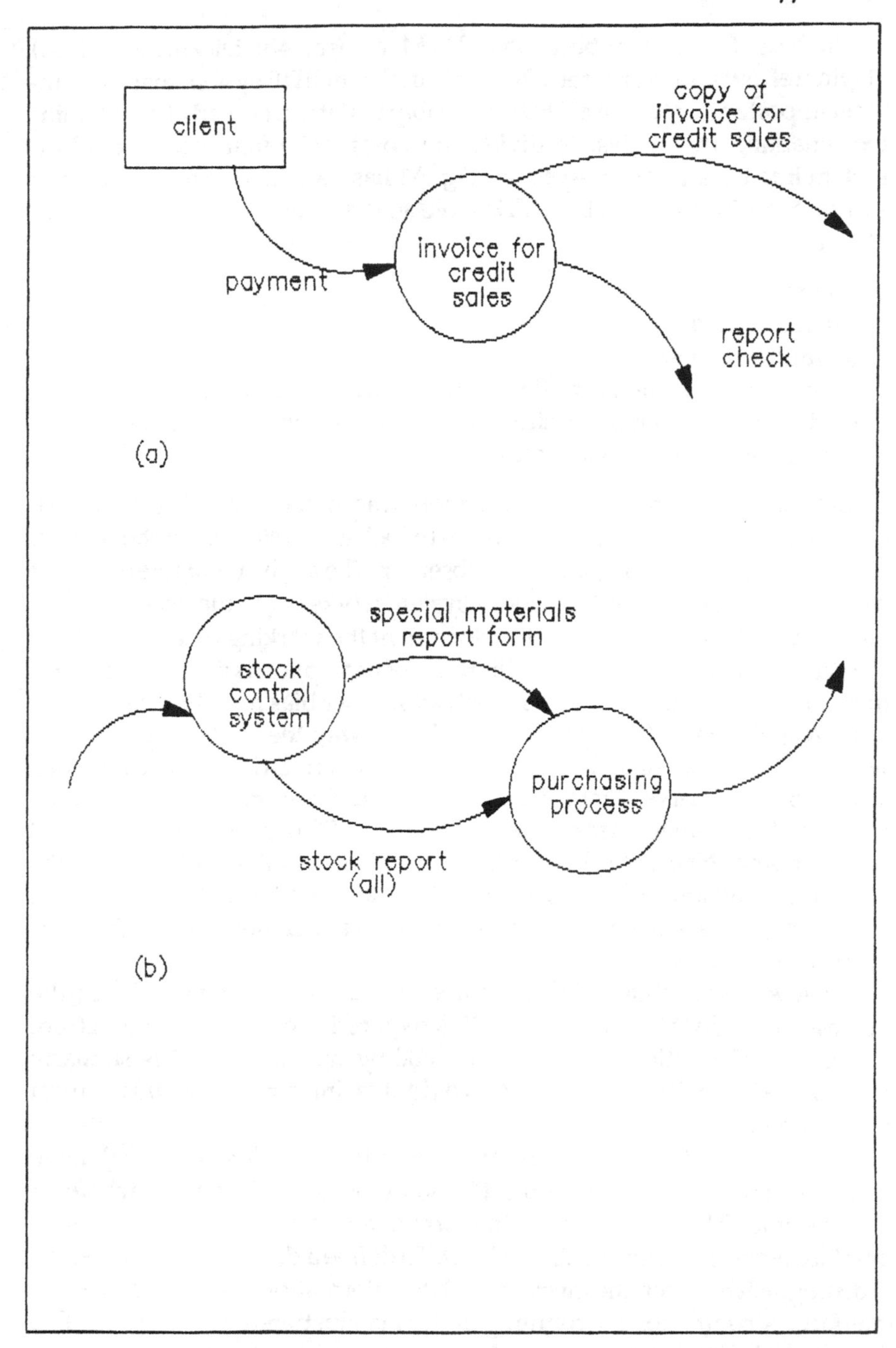
client
copy of
invoice for
credit sales
invoice for
credit
sales
payment
report
check
(a)
special materials
report form
stock
control
system
purchasing
process
stock report
(all)
(b)

Fig. A3

The best of the key publications – De Marco (ref. 48), Dickinson (ref. 58), Flavin (ref. 59) and Ward (ref. 60) – confirm that in EDP system mapping, the most important feature of a DFD is probably that it is a powerful partitioning tool enabling the analyst to divide the observed, often complex subject system into its natural sub-systems. Fig. A1 has been broken down into nine sub-systems by using DFDs. DFDs have several significant characteristics. They are

(a) graphical
(b) partitioned
(c) multi-dimensional
(d) designed to emphasize flow of data rather than control
(e) designed to represent a situation from the viewpoint of the data rather than a person or an organization.

Jackson (ref. 47), representing the more traditional view of systems analysis, suggests that the analyst attempts to look at a system from the point of view of the user, a participant or an observer. The analyst will interview the user to find out how a system or sub-system works (from the user's point of view). After this he will attempt to document the working of a system from the system's point of view. This difference of perception is unsatisfactory for management system research as it will allow user bias to influence results.

SDA is different. It maps in a rule-driven way the route or path along which the observed data flow. It also records in a rule-driven way the exact format of the observed data at each stage. It is therefore verifiable at each point, and for management system researchers SDA provides a consistent basis for describing data flows within organizations. An analyst using the SDA technique attaches himself to individual pieces of data, and follows each through the system, in a precise and logical manner. The symbols are illustrated in Fig. A1.

In Fig. A2, D1 is obtained from source S1 and transformed into D2 by the process P1. To do this, access to file F1 is required to draw information from it. Further information is drawn from P30 by way of D9. D2 is similarly transformed into D3 by process P2. To do this, information is drawn from P11 by way of D6.

The processes P1 and P2 receive their name from the data flows that move into and out of them. For example, P2 should be given the name 'transform D2 + D6 into D3'. A data flow in the form of a named vector represents an interface between components of a DFD. To define a data flow it is necessary to distinguish between the interface and the information that passes over the interface. A contractor's accounting system is illustrated in Fig. A3(a). The data flow labelled 'payments' consists generally of the client's 'pink copy of an invoice' and 'a cheque'. This 'packet' of information is composed of 'a

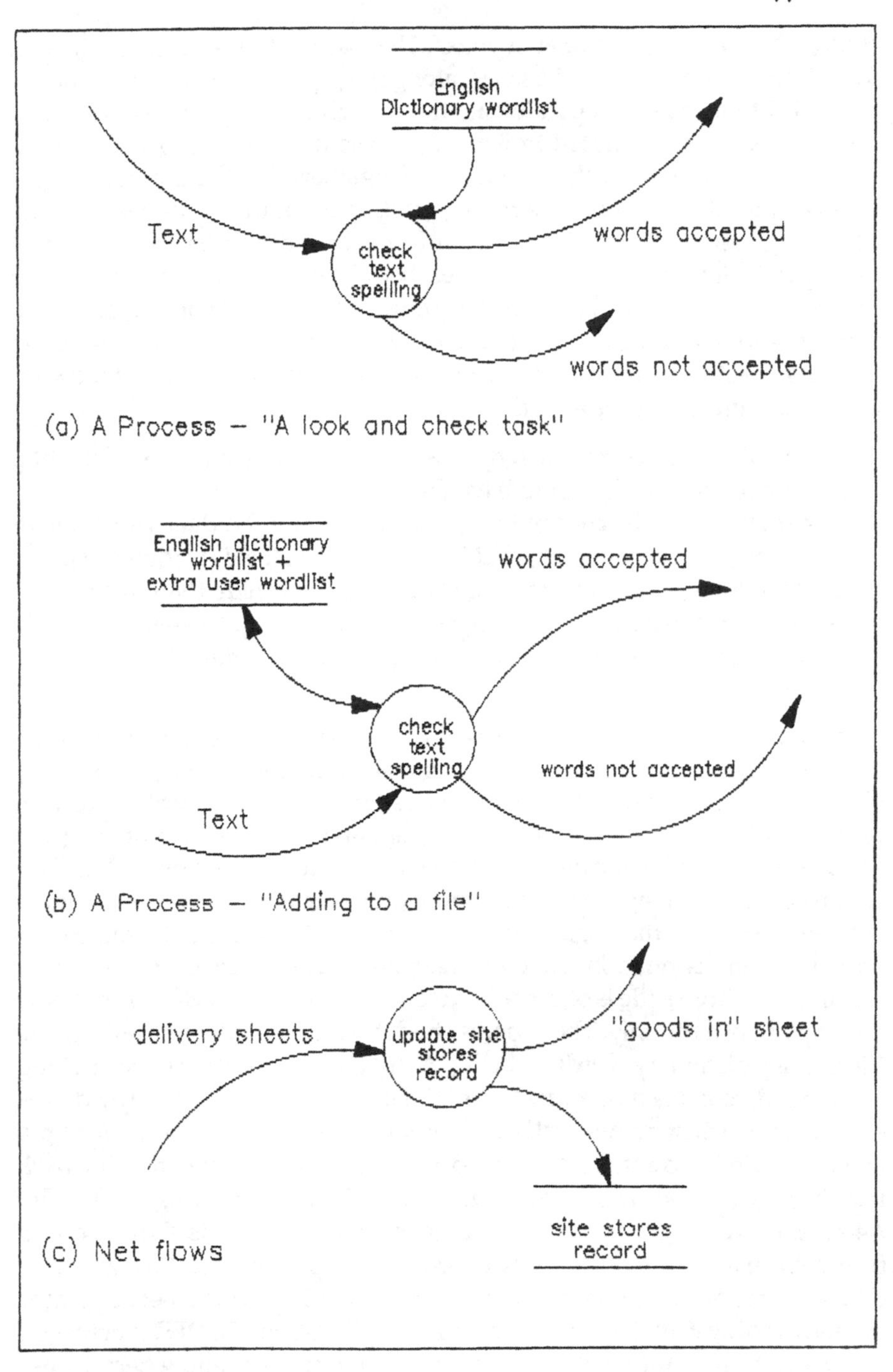
English
Dictionary wordlist
Text
check
text
spelling
words accepted
words not accepted
(a) A Process — "A look and check task"
English dictionary
wordlist +
extra user wordlist
words accepted
check
text
spelling
words not accepted
Text
(b) A Process — "Adding to a file"
delivery sheets
update site
stores
record
"goods in" sheet
site stores
record
(c) Net flows

Fig. A4

cheque' and 'a pink copy of an invoice'. The 'packet' flows from the client to the process 'invoice for credit sales', along the pipeline called 'payment'.

A DFD that has two separate data flows, each moving between the same two processes, is illustrated in Fig. A3(b). The different items do not constitute a 'packet' because they never travel together. 'Stock report (all)' might be a weekly site-to-head office report, while 'special materials report form' could be generated immediately and whenever the prescribed re-order level for a particular critical material is reached. The data in two pieces have different purposes, could possibly be provided by different bits of the source process, and have little to do with each other. They are classified as separate 'packets', each having its own precise and relevant name. The following notational conventions are used

(a) data flow names may be hyphenated and written in the form of a title
(b) no two data flows should have the same name
(c) names should be chosen to represent not only the data which move along the pipeline, but also what is known about the data themselves
(d) data flows that either converge or diverge can share the same name
(e) data flows that move into or out of files do not require names. The file name is enough to describe the data flow, all other data flows must be named.

Another part of the DFD is the process which identifies or represents some kind of work performed on data, and is given a unique number. Fig. A4(a) shows a 'look and checks' task that divides the incoming flow of words into two 'packets'; words accepted (correctly spelled), and words not accepted (incorrectly spelled or not on the dictionary list). Thus an incoming data flow is transformed into separate and different outgoing data flows. Clearly there is a need to specify the English dictionary wordlist in Fig. 4(a) by more than its name. This is done in the data dictionary. The direction of an arrow leading to or from a file is of considerable importance. Fig. A4(a) shows data moving out of a file only. The process check text spelling does not add to the file English dictionary wordlist. In Fig. A4(b) the process check text spelling is modified, to make new entries on to the file English dictionary wordlist + extra user wordlist, under certain conditions. Fig. A4(c) shows an example of data flowing into a 'site store record file'. No data are shown flowing back into the process. The process cannot update without first reading the file. Fig. A4(c) is, however, correctly drawn because only net flow is shown to and from a file. If input is required only to produce output, then net flow is clearly out. A source or sink (or terminator) illustrated in Fig. A1 is a net originator or receiver of system data, thus dramatic simplification of a DFD is achieved by showing where the net inputs to the system come from and where the net outputs from the system go. Sources and sinks (such as people and organiz-

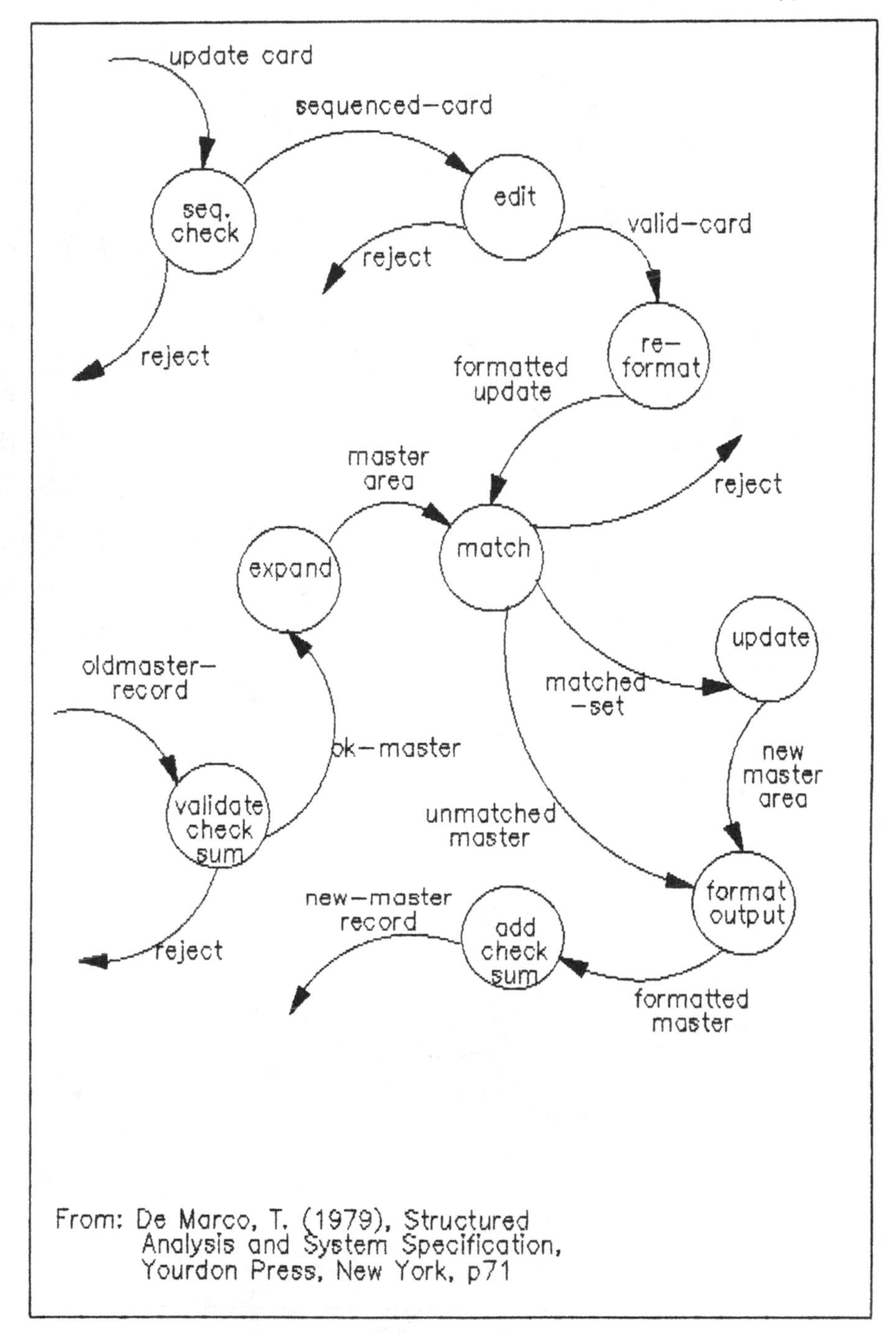

Fig. A5. The need for levelling data flow diagrams

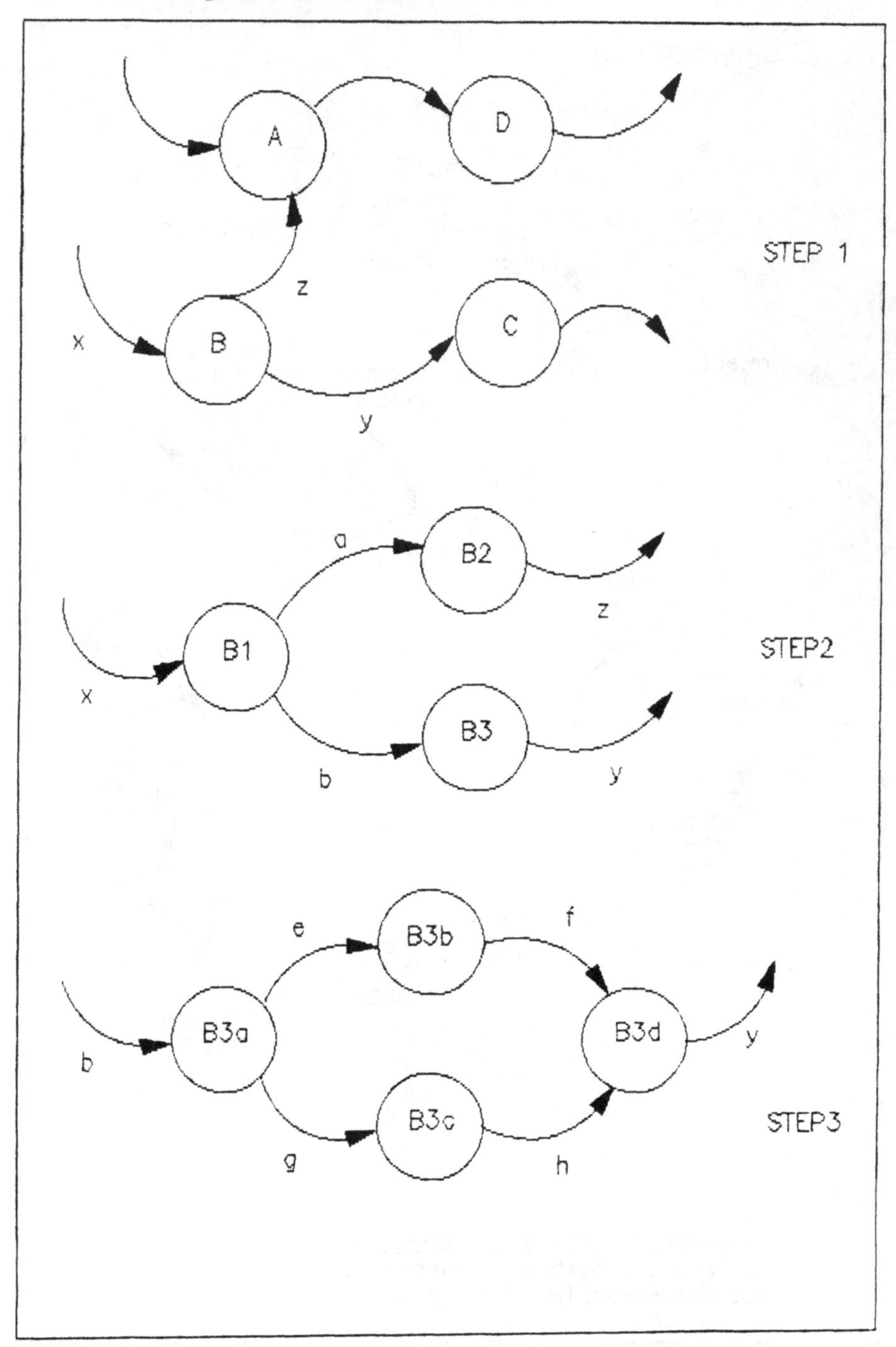

Fig. A6. Partitioning into sub-systems

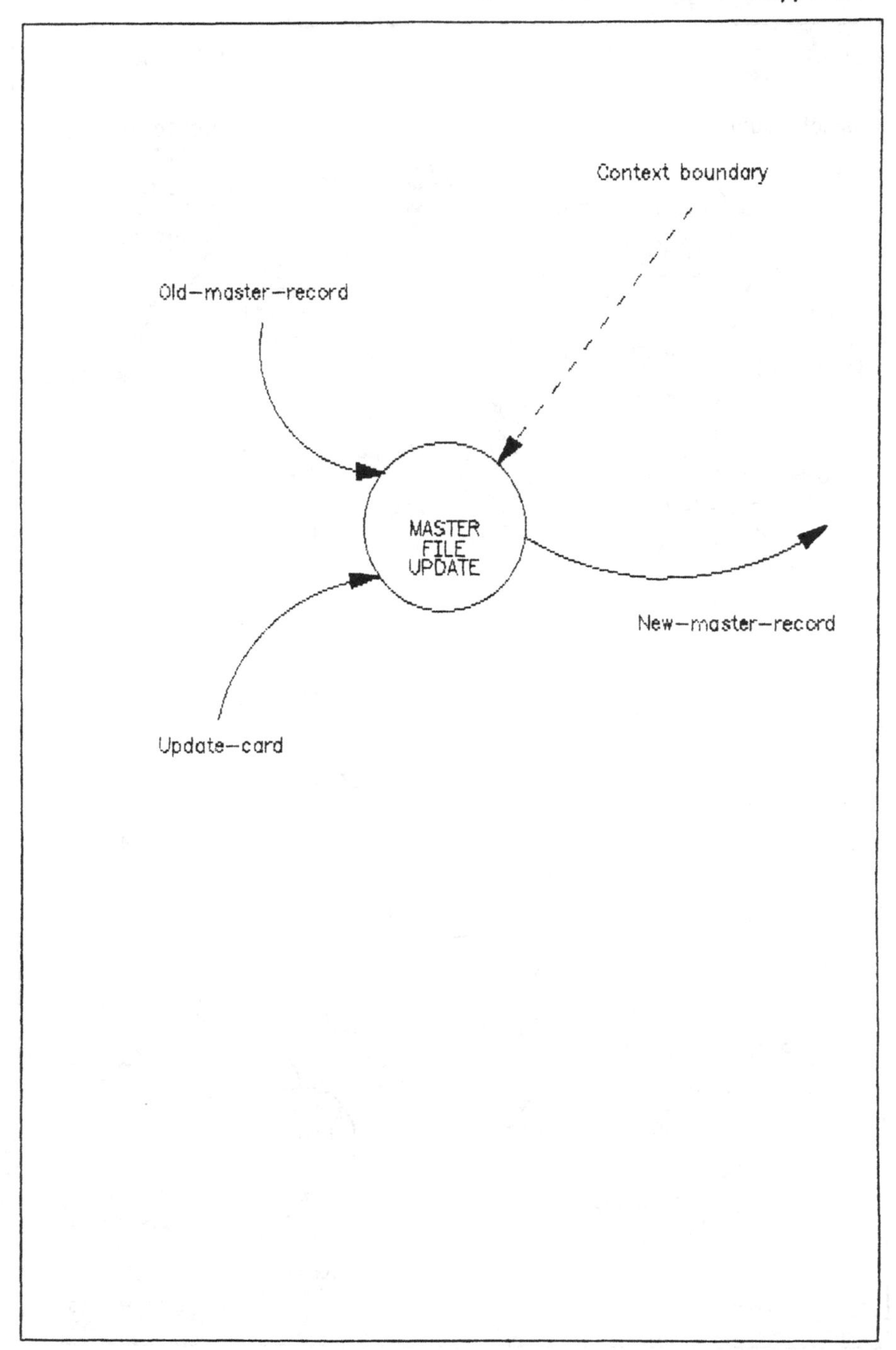

Fig. A7. Context diagram

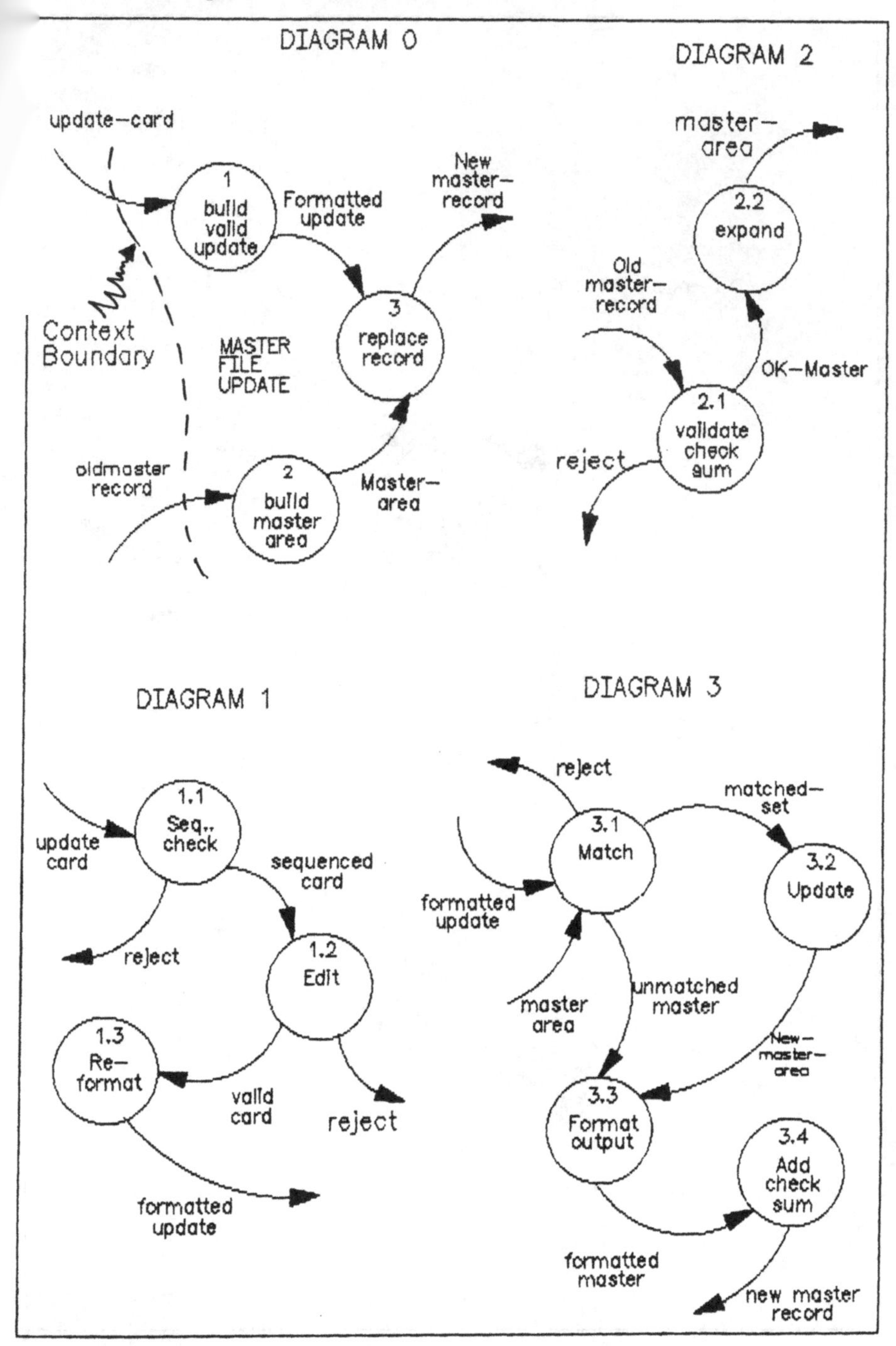

Fig. A8. A set of levelled DFDs

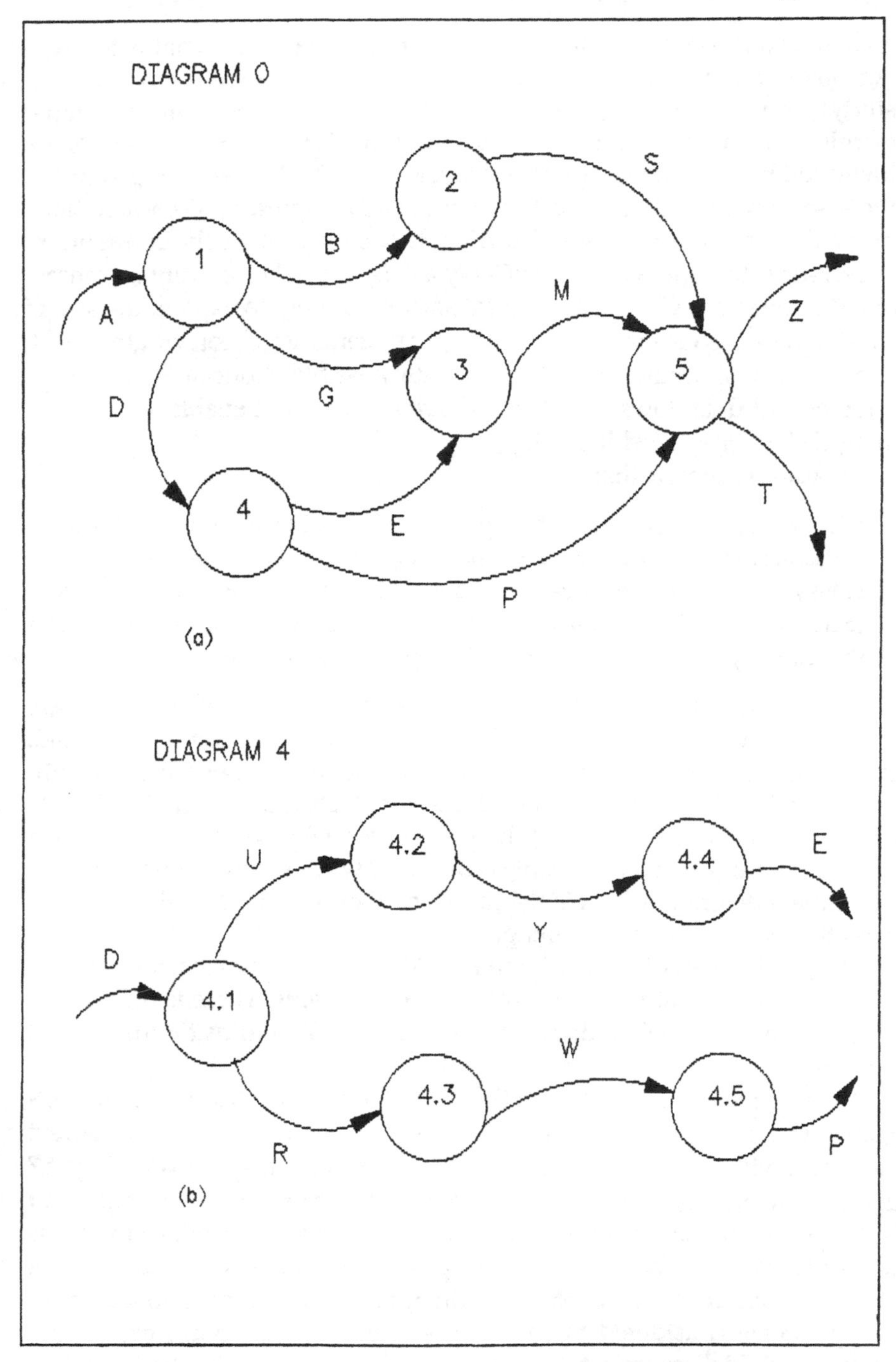

Fig. A9. Balancing DFD sets

ations) lie outside the 'context' or boundary of the system. That is to say, a person or organization lying within the context of a system (the domain of a study) would be characterized by the process he or it performs. A number of rules for understanding SDA and, in particular, for drawing DFDs are identified by De Marco (ref. 48) and others (refs 58-60). The rules governing the system partitioning process are particularly important. De Marco illustrates this problem well (see Fig. A5). He shows a relatively unimportant sub-system, but one which is probably as big as can be usefully accommodated in one unit. Clearly it is not satisfactory merely to expand the size of the diagram to cope with larger and larger systems. Precision would be lost, and an analyst would have to cope with possibly thousands of process bubbles and data flows. De Marco identifies rules that enable a top-down analysis known as levelling or layering.

De Marco suggests that

> when a system is too large to be represented by a DFD on a single A4 sheet, it is necessary to undertake an initial partitioning into major sub-systems. If the sub-systems are still too large they should in turn be divided into sub-sub-systems and so on. In this way an analyst will end up with system components that can be portrayed by simple DFDs of primitive functions.

Primitive functions or functional primitives, as they are better known, are the lowest level component of a DFD. They cannot be further decomposed. If each division of the system is partitioned into sub-systems and sub-sub-systems with a DFD for each, a levelled set of DFDs is obtained. Figs A6-A8, show part of a levelled DFD set. In any levelled DFD set there should be a number of important recognizable elements. Two of the most important are the context diagram, which is the top level diagram, and functional primitives that defy further partitioning.

The processes in diagrams 1-3 in Fig. A8 are all functional primitives. In any DFD set there are in effect two context diagrams. There is the context diagram and the level '0' diagram which essentially builds on the context diagram.

A context diagram looks from the subject system context boundary *outwards*, and the level 0 diagram looks from the subject system context boundary with its environment *inwards* at the major sub-system (see Figs A7 and A8). In Fig. A6, the level 0 diagram would be represented by step 1 and is the 'parent' of the first level breakdown diagram (step 2), which in turn is parent to its own 'child diagram' (step 3). This process continues until a functional primitive is reached. It should be noted that the partitioned layers in Fig. A6 are considered to be 'balanced', because all data flows shown entering a child diagram must be represented on the parent by the same data flow into the associated bubbles. All the outputs from the child diagram must

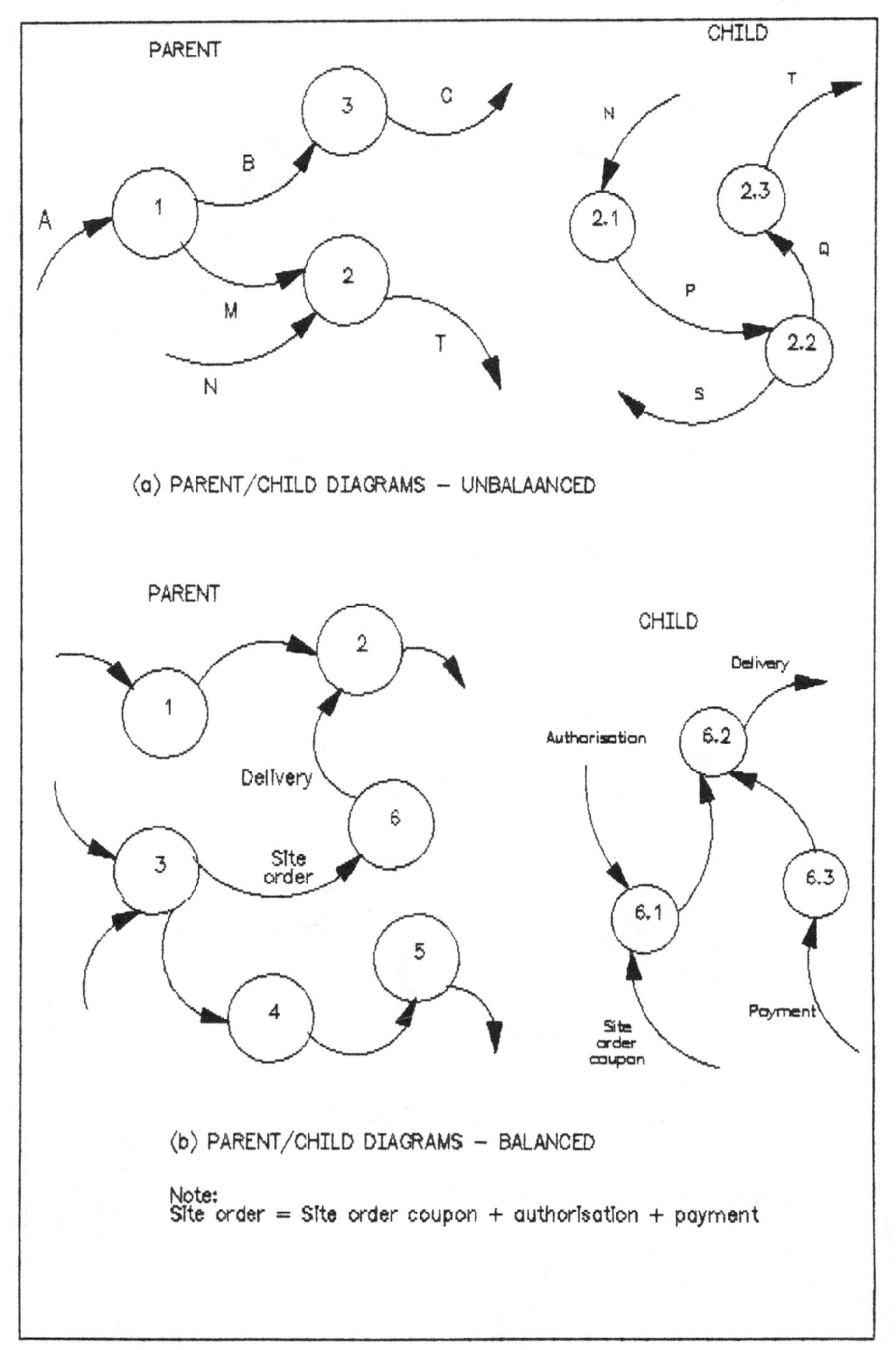

Fig. A10. Balancing DFD sets: parent/child relationship

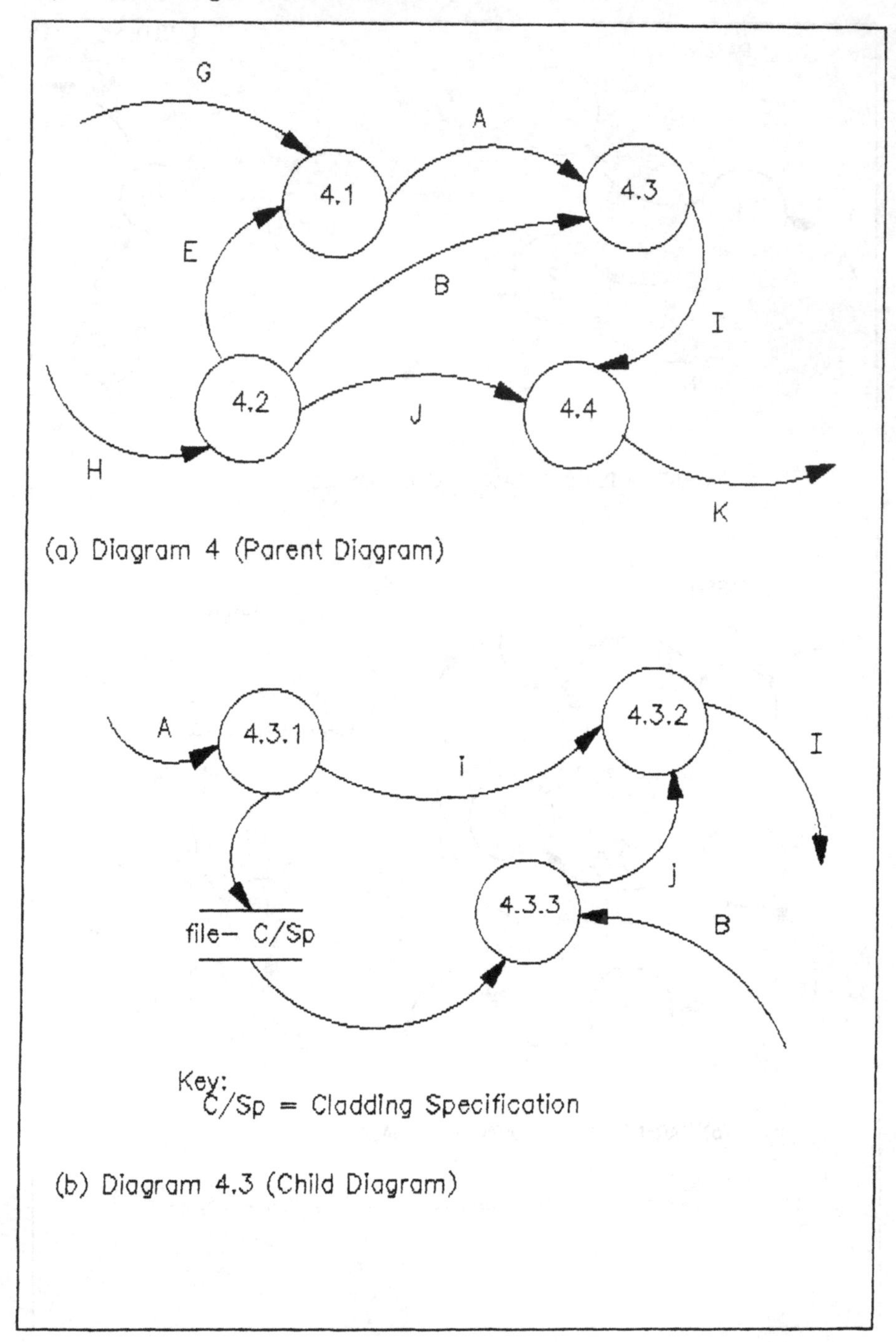

Fig. A11. Primitives

be the same as outputs from the associated bubble on the parent. For example, process step 1 has data flow X entering it, and data flows Y and Z leaving it. In step 2, process B1 has data flow X entering it, and data flow Y leaving process B3 and data flow Z leaving process B2. The same balancing exists between steps 2 and 3 as process B3 is partitioned. Figs A9 and A10 give further examples of balancing. In Fig. A9, process 4 is partitioned (see Fig. A9(b)) into five sub-systems, but is balanced because the one input D and the two outputs P and E balance. Fig. A10(a) shows a typical example of an unbalanced partitioning, while Fig. A10(b) shows a balanced partitioned example from a contractor's site management system. To ensure consistency of approach when a system is being mapped, a clear and precise description is needed at the functional primitive stage. In Fig.11 it can be seen that each of the processes in diagram (b) is a functional primitive because each one has been partitioned to the point where there are five or fewer processes entering and leaving, and where the nature of data in those processes defies further useful reduction. Clearly they must be documented, and in some way other than with a DFD description (as there is no lower level). To do this a mini-spec is used for each DFD bubble which is not further decomposed, i.e. for each functional primitive. Mini-specs should be identified by the number of the related process. Clearly if a mini-spec were written for process 4.3 on the parent diagram (diagram 4 of Fig. A11), it would have to be a duplication of the total of the individual mini-specs on the diagram 4.3 (Fig. A11(b)), as 4.3 (the parent) would have been completely specified. It should be emphasized that 4.3 on the parent is exactly equivalent to the whole of diagram 4.3.

Structured walkthrough (see note 1) provides a rigorous check against conceptual errorat each stage of the partitioning.

The data dictionary is part of the structured specification. De Marco (ref. 48) states that there is one DD entry for each unique data flow that appears anywhere in the DFD set. There is one DD entry for each file referenced on any diagram in the set. There are four classes of data to be defined in any DD: a data flow, a file, a process, and data bases such as sources and sinks.

It should be clearly understood that most definitions in a DD represent the item being defined as a combination of components. The components are in turn defined as being composed of lower level components and so on. Thus the DD also follows a top-down partitioning methodology. Definitions in the DD are therefore top-down partitioning of data. If it is known, for example, that data flow 1 is composed of 2, 3, 4 and 5; that 2 is composed of 2.1, 2.2, 2.3; and 3 is composed of 3.1 and 3.2; that 4 is composed of 4.1, 4.2, 4.3 and 4.4; and that 5 cannot be partitioned, then 1 = [2.1 + 2.2 + 2.3] + [3.1 + 3.2] + [4.1 + 4.2 + 4.3 + 4.4] + 5. From this it can be seen that with complex data flows this approach would reduce readability (i.e clarity). It is necessary therefore to define data in terms of high level subordinates, and then in turn

Table 1. An example of a data dictionary entry for part of a construction project management system – as made by an EDP systems analyst

contracts	= main contract + {nominated sub-contracts}
detail brief	= {consultant/client questionnaire answers} + {client briefing answers}
fee invoices	= [{percentage based fee invoice} {time based fee invoice}]
initial requirements	= initial statement of requirements + {client answers}
invoice line	= quantity + item number + unit price + item subtotal
monthly costs	= {month + job number + {staff names + hours}}
offer of services	= description of proposed profession services + fee for: [project management + architectural design + building engineering design site supervision]
project brief	= initial requirements + decision to proceed + detail brief
staff allocation	= [{site staff name + role} {design staff name + role}]
site staff requirement	= specification for resident engineers + clerk of works

to define those subordinates. Thus the definition will move logically from the most abstract to the most detailed. Almost all DD definitions are formulae that declare that which is being analysed, recorded or designed, as being made up of a related set of components.

The following five notations describe five relational operators (see note 2)

(a) = IS EQUIVALENT TO
(b) + AND

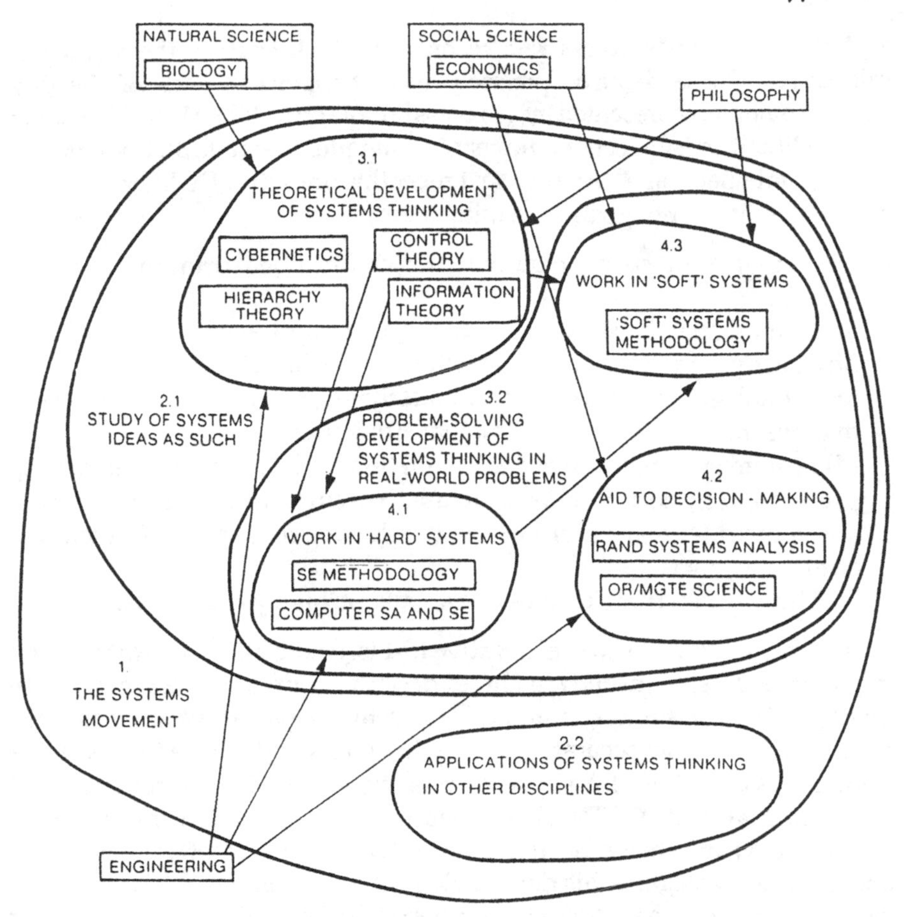

Fig. A12. Shape of systems movement, indicating major external influences (from : Checkland P. B. Systems thinking, systems practice. Wiley, Chichester, 1981, 97)

(c) [] EITHER/OR; (i.e. select one of the operations enclosed in the brackets)
(d) {} ITERATIONS OF the enclosed component
(e) () the enclosed component is OPTIONAL.

Table 1 provides an example of a DD entry. Although partitioning by both the DFD and the DD is a vitally important activity it would be quite unsatisfactory to specify by partitioning alone. The make-up of the pieces needs to be recorded, and both the system and the data need also to be actively broken down into finer and finer pieces.

An analyst needs access to a technique which enables the writing of mini-specs. These describe precisely what happens in each of the data transformations represented by a process bubble on a DFD. The EDP systems analyst literature suggests a number of techniques (see note 3). Both process notes (mini-specs) and narrative text meet this objective. De Marco (ref. 48) suggests a number of goals for a mini-spec.

(a) A single mini-spec is required for each functional primitive in the set of DFDs.
(b) A mini-spec must describe rules governing transformation of data flows arriving at the associated primitive into data flows leaving it.
(c) A mini spec must describe an underlying policy governing transformations, but not a method of implementing that policy.
(d) A mini-spec must state transformation policy, without introducing overlap of any sort into the structured specification, i.e. each must describe the rules that govern the way inflowing data is transformed into outflowing data.
(e) The way of writing a mini-spec should be highly orthogonal.

Another approach is to use narrative text (see note 3). This is widely used in the broader 'systems movement' to describe or understand what is happening within a subject system. Clearly the better a system has been partitioned, the more accurate and easy to understand will be the specification. Checkland (ref. 46) supports this view with his concept of 'root definitions' and his CATWOE methodology. However, an inspection of Checkland's work shows that it is designed primarily for 'soft system' work, and at the early stages of his methodology is the requirement to develop an expression not of the problem to be considered, but rather 'the situation or environment surrounding it...'. Also, the Checkland methodology again at an early stage requires the naming of some systems that 'look as though they might be relevant...', and preparing concise definitions of what these systems are. With the CATWOE approach, it is only at a late stage that the (conceptual) models are brought into the real world and set against a perception of what exists. This approach has clear benefits in the area of soft systems (see Fig. A12), but from the point of view of hard systems it can offer little more than an impression of a situation. By contrast, the technique developed by EDP systems analysts for hard systems and SDA is based on initial precise specifications of the form and content of data recorded in the DD with the narrative text and mini-specs. The least rigorous aspect is undertaken last, thus minimizing bias. With the Checkland approach the least rigorous aspects are undertaken first (see note 3) (ref. 46). This contrasts dramatically with the SDA approach, where any researcher bias could severely influence results. So while the SDA approach is preferable for research, its mini-spec or narrative text aspect is its major weakness in terms of rigour, and this needs further work. Some useful improvements have been suggested by Dickinson who advocates that 'policy logic' of a process can be recorded precisely by means of 'decision tables' and 'decision trees' (ref. 58).

Necessary changes for management systems research

Fieldwork undertaken over the last eight years in the construction industry would suggest that improvements can be made to the technique to make it more suitable for management systems research. The techniques described earlier in this chapter have been carefully selected from the full suite of EDP and SDA techniques as the most relevant to management systems research.

Account should be taken of the fact that EDP systems analysts and management systems researchers have very different objectives. According to Kefalasand and Schoderbek (ref. 61), the objective of an EDP systems analyst would be 'organized step by step study of the detailed (usually manual) procedures for the collection, manipulation and evaluation of data about an organization for the purpose, not only of determining what must be done, but also of ascertaining the best way to improve the functioning (and automation) of the system'. While Medawar (ref. 34) suggests that the objective of the researcher would be the 'undertaking of exploratory activity, the purpose of which is to come to a better understanding of the natural world or of some phenomena...'.

For a management system researcher the words 'natural world or of some phenomena' should be replaced by 'management systems'

Despite clear differences between the two definitions, commonality does exist in the undertaking of exploratory activity, or step by step study of detailed procedures for the collection, manipulation and evaluation of data about an organization, i.e. the activity of mapping the observed or existing management systems. This point is an important guide in the application of SDA to management systems research.

The purpose of the SDA tools can usefully be divided into three groups.

- *Group 1.* The data and logic of its flow; those that enable the partitioning and documenting of flows of information.
- *Group 2.* The recording and documenting of the interfaces between partitioned sub-systems; those that allow the analyst to keep track of and evaluate interfaces between partitioned sub-systems.
- *Group 3.* The writing of mini-specs to describe processes including functional primitives; those that enable or assist in the preparation of the mini-specs necessary to describe the functional primitives.

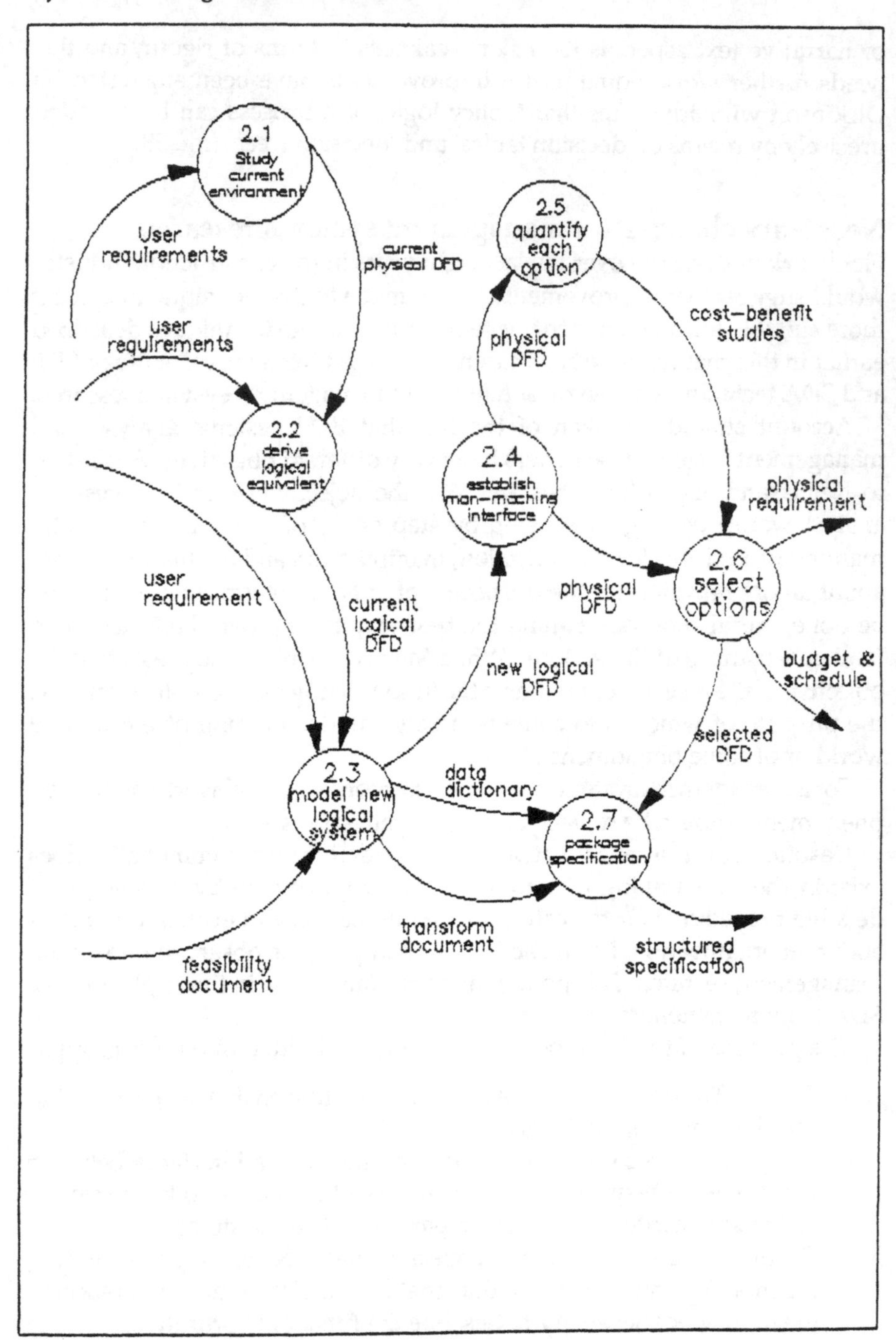

Fig. A13. SDA technique as used by EDP systems analysts

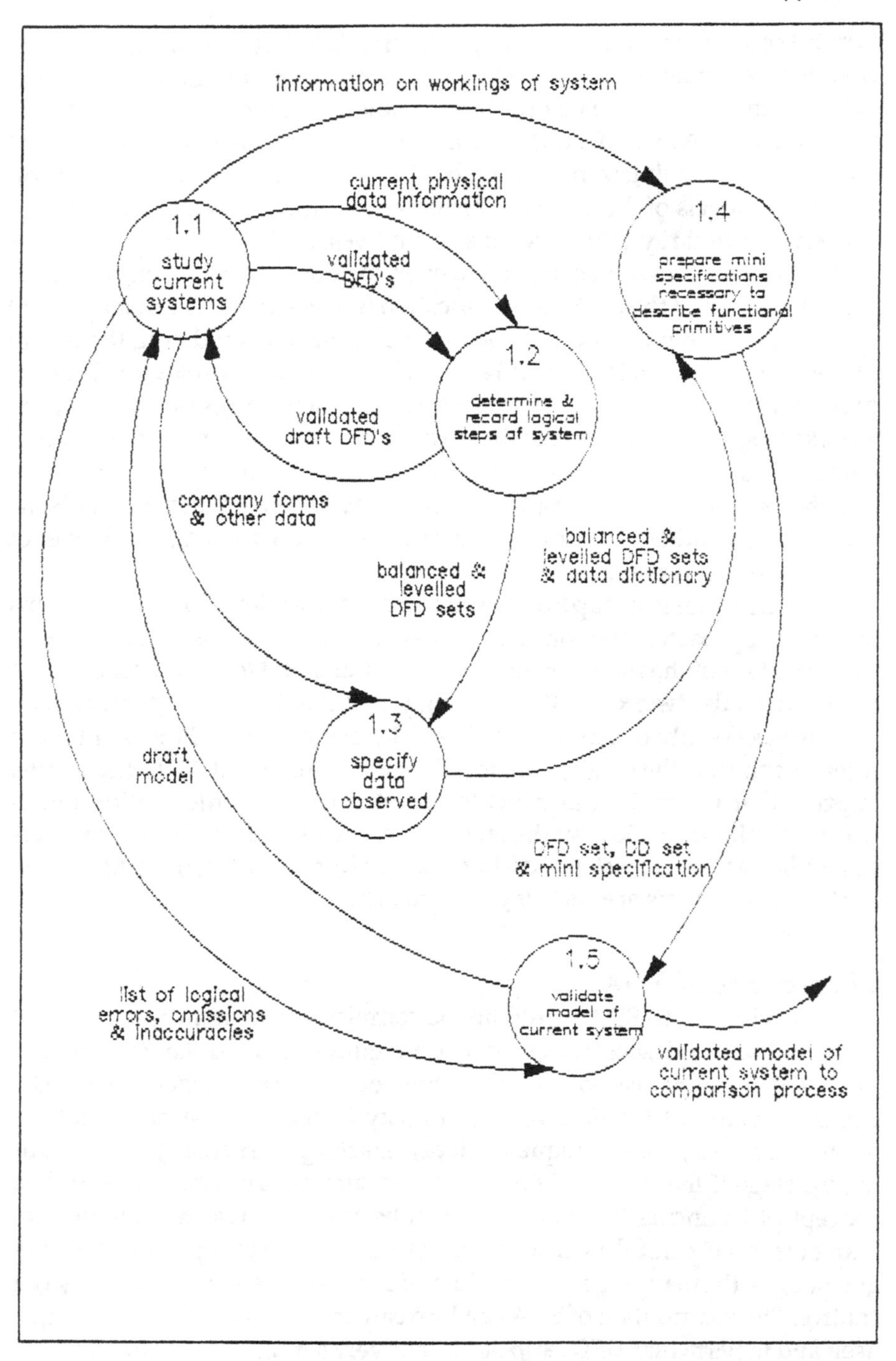

Fig. A14. SDA technique as used by management systems researchers

Careful consideration supported by extensive field testing has indicated that changes are not necessary to the techniques in group 1 but rather to the way in which they are used. The EDP systems analyst approach needs modification (see Figs A13 and A14), not in terms of system partitioning, DFD construction levelling or balancing, but in the need for greater emphasis on the middle levels of the system. This is because the EDP systems analyst is concerned primarily with functional primitives, as they make up the major part of the functional specification needed to allow him to develop the new logical model and the package specification (see Fig. A14). The management systems researcher is concerned with developing and validating the model of the current system (see Fig. A14). For the EDP systems analyst the model of the current system is only one necessary step in the process of developing a package specification, but for the management researcher it is the object of study. This difference in approach is reflected in the management researcher's need to give the middle levels of the system greater emphasis; he must understand and record the current system at each stage and level of the partitioning process.

The techniques in group 2 also need some modifications. The EDP systems analyst's approach to the construction of a data dictionary is consistent with the objectives of the management system researcher. However, conclusions drawn from fieldwork results have suggested that the EDP approach may have unnecessarily different objectives. The higher level DFDs are of more interest and thus the concept of narrative text more useful. Field test results suggest that group 2 does provide useful contextual information but is necessarily less reliable than the other techniques, because of possible investigator bias and inconsistency owing to the lack of rule-driven methodology. Further work in this area is being undertaken.

Limitations of SDA

This introduction to SDA would not be complete without an assessment of its limitations and weaknesses, and some guidance as to the level of confidence that can be placed in results obtained from the methodology. SDA has been found at the fieldwork and analysis stages to be highly labour intensive. In use, the technique is highly exacting, particularly at the writing-up stage if the rules and procedures as summarized are followed. The concept of balancing the data flows can be taxing in reality. Potential researchers need careful instruction in its use if the concept is to be fully grasped, i.e. the mapping and recording of data flows rather than managerial control. The contribution of SDA can be easily missed because the technique itself and in particular DFDs appear deceptively simple in their presentation and use at a superficial level.

It should be emphasized that, so far, no fieldwork has been constructed or undertaken to test the level of consistency of the technique's results: i.e. recording the same thing a large number of times, and measuring the level of consistency of the results. The fact that the technique is based largely on the recording of hard data, that it is primarily rule driven, and that the least rule-driven aspect is used last where any investigator bias will have a minimum effect, suggests that the level of consistency would be sufficient to warrant its use and further development. Another limitation of the SDA technique is that it takes no account of managerial control or responsibility nor of human, cultural or other socio-economic aspects. Any systems map that is produced is therefore limited to flows of data and, as such, in managerial terms is looking only at one important aspect.

Conclusion

The design and recording of data flow systems will become increasingly more important, and despite its limitations, SDA is capable of providing a consistent basis for describing the flow of information, both formal and informal, within an organization. It is its ability to allow comparisons that suggests a new and potentially fruitful avenue of research as computer-integrated construction becomes inevitable.

Notes

1. EDP systems analysts have split the concept and practice of structured walkthrough into two types

(a) those at the analysis phase (refs 38, 48, 51, 60, 62)
(b) those at the implementation phase.

2. Bohm and Jacopini (ref. 63) show that any process or programme can only be made up of elements that are related in one of three ways

(a) sequences: the linking together of two or more components in order
(b) selection: the choice of one of two or more possibilities
(c) iteration: the repetition of a designated component any number of times.

Two further elements have been added for greater clarity and ease of use

(d) is equivalent to: in effect, the same as
(e) optional: zero or one iteration of a particular component.

For a detailed explanation see De Marco (ref. 48).

3. An analysis of the literature would suggest five ways of successfully describing and specifying a primitive: narrative text, process notes (mini-specs), structured English, decision trees and decision tables (refs 48, 51, 58, 60, 62, 64). These references also describe a consensus view of the rules governing the writing of mini-specs.

Of the five methods, only narrative text and process notes are discussed, as only they meet fully the objective of enabling the writing of mini-specs which describe precisely what happens in each of the data transformations represented by a process bubble on a DFD. For a contrasting view, one developed primarily for 'soft systems research', see Checkland (ref. 46), and the work of the Open University Systems Group between 1981 and 1989 (ref. 65).

Appendix 2. Some reflections on the robustness of the model and on future directions

The conclusions that can reasonably be drawn are best examined in two parts. First, there are the conclusions specific to the study, or what can be learned about the systems analysed, and about SDA when used in the kind of fieldwork described in chapter 1. Second, there are the general conclusions, that look from a more detached level and in a less focused way at where this sort of research is heading, and where other benefits can be gained, and more general conclusions drawn, if this approach is applied.

Conclusions specific to the study

The model described in chapters 3 and 4 represents a 'map' and a specification of the processes and the data flows within and around a typical contractor's management system for a major construction project. It represents an inter-company information system study and is a GDFM. It is a general system that combines the best of the systems observed in three major UK contractors, i.e. the common aspects of all three companies, and where there is significant system divergence, the best aspects drawn from each of the three companies. Clear evidence was found during both the fieldwork and the analysis stages to suggest that all three company systems had evolved in an incremental *ad hoc* way. There was also evidence to suggest that none of the three companies had looked at, or attempted to improve their systems as a complete entity, despite considerable system defects and redundancies that were fairly obvious when the system was looked at and in a holistic way. There was considerable evidence that many parts of the individual company systems, particularly cost-related sub-systems, had been repeatedly changed, amended, or added to without any thought for the effect of the changes on the system as a whole. There was little evidence that there had been any attempt by senior managers to stand back and view the system as a whole, or the whole system as a 'management system for a major construction project', or to simplify systems added to repeatedly and, as a result, made more complex.

A provisional explanation for specific commonality and variance is based on the system data and the contextual data gathered for each of the companies. It would appear that data flow systems are affected by the flexibility of the company. This includes the ability of the company to respond, its speed of response, and the general awareness of its senior managers (i.e. not over-focused) to

(a) the way in which the company's business is acquired and undertaken
(b) the socio-economic environment within which the company operates
(c) the history of the company, including the nature and skills of the founder, the original or previous business of the company, and the desire and the will of its most senior staff not to become or remain a prisoner to that history
(d) the level of education, the breadth of vision, and the dominant professional background and culture of senior and other key staff in the company.

Of the three companies looked at, 'systems evolution' was clearly visible in all three, and 'cultural lag' in one of the two most system-biased companies, i.e. the one with the production bias (see Tables 2 and 3). It is possible from the data to suggest that each company was showing a distinct bias in its data system. These biases were to

(a) production
(b) cost
(c) performance (a performance-oriented system generally well-balanced between (a) and (b).

Clearly these suggestions represent only provisional classifications and a great deal more work is necessary if they are to be proposed with any confidence.

The model as outlined in chapters 3 and 4 does, however, provide a clear basis for comparison. It is research work that stands in the public knowledge domain for senior managers and systems analysts to compare with their existing company system. For this and other purposes it forms the first steps towards a yardstick, against which general improvements can be made and measured, much like the CIOB's *Code of estimating practice*, although this was on a more limited scale (ref. 28).

The publication of this model also begins to demystify the considerable managerial skill and complex systems necessary for successfully managing construction. By specifying clearly and in a structured way the data system behind a contractor's management system for a major construction project – the back-up that supports and prompts a successful manager – this can be achieved.

As an analogy, personal medical hygiene does not ensure health, but it will reduce the likelihood of becoming ill. In much the same way a contractor's manager managing a construction project will not be successful simply because he has an appropriate data system supporting him. It will also require personal skills and qualities, managerial ability, appropriate experience, and technical and specific project knowledge. However, an appropriate supporting data system will encourage, for example, better decision making, and will reduce the likelihood of managerial failure. This is good 'project hygiene', with appropriate well-designed hygiene factors particularly in the light of the amount of time this type of manager spends dealing with information (see chapter 2).

It is also very important that such a model will aid practice, teaching and training. It also goes some way towards developing the algorithm behind any future training simulations for a contractor's project management. But

Table 2. Type of control system bias identified in sample companies

Company	Progress and Productivity	Cost	Performance
A	▪		
B		▪	
C			▪

Progress and Productivity = [long/medium/short term scheduling and productivity indicators + method statements + incentive schemes + sub-contractor co-ordination and control]

Cost = [external valuation + cost value comparison procedure +internal costing + material purchase + subcontractor selection + subcontractor payment + subcontractor cost control]

Performance = [broad balance between both "progress and productivity" and "cost", in terms of overall data flow within the management monitoring and control system]

Table 3. A summary of the divergence and imbalance between the systems of companies A, B and C

COMPANY	SYSTEM WEAKNESSES*	SYSTEM STRENGTHS*
A	.contract budgeting system .internal valuation system .contract budget monitoring .cost value reconciliation system .cash flow monitoring system	.information management from the designer - at tender stage .construction management reporting .site management control procedures .management system for directly employed labor
B	.information management from the designers - at tender stage .site management control procedures .construction management reporting .management system for directly employed labor	.programming the tender system .contract conditions and risk assessment .cost value comparison system .internal valuation system .contract budget system .contract budget monitoring .cash flow monitoring system

* Weaknesses and strengths are relative here and are as a result of a comparison with the system of company C. Company C has been used as the 'bench-mark' against which the other two have been compared.

possibly more important, from the point of view of the development of the discipline, is that this model and the SDA technique together provide a piece of work and a research methodology which others in the UK, EC, and international research community will be able to challenge, refute, amend and hopefully to develop and build on. It will be of immense value to be able to analyse the GDFMs, and in time the structured data systems maps (SDSMs), that flow from similar studies conducted on a range of different types of contractors, designers and client's project managers, in the US, Japan, and other EC countries, in particular France, Italy and Holland. This research has shown that when the rule-driven SDA technique is used as the research tool for systems studies of data flows, this sort of work is possible and useful.

General overall conclusions

Looking to the future, SDA and the various data models produced might have possible uses and benefits in other areas of IT: for example, in the redesigning of products, processes and procedures. EDI is an obvious area, and one that is likely in time to change quite radically the flow of data both inside and outside a company. Other relevant candidates are executive information systems and the concept of total 'computer-integrated construction', which is considered at the end of this Appendix.

The GDFM that has been developed as a result of this study appears to have important implications for expert systems, robotics, the building process and the holistic design of management information systems. At a more ambitious level, it possibly has important implications for the design of integrated management EDP systems for computer-integrated construction.

Expert systems

This study has contributed to attempts at linking into one SDSM all DFD functional primitives of one sub-system of the construction process. For example, the GDFM of the construction management system used by major UK contractors on large commercial building projects and developed as a result of this research project could be linked with another GDFM, such as 'the project design process', developed as a result of other research. Other recent research at the University of Reading has demonstrated this to be both possible and successful (see chapter 1, note 2). This is because the SDA technique accurately identifies and specifies the interface points between any two sub-systems of an overall data system behind, for example, the procurement of a major construction project. By building up an SDSM of the total construction process in this manner and showing in a structured way the flows of data and the process points where data are manipulated or processed in some way, existing stand-alone expert systems can be plotted on to the SDSM and the interfaces between them specified. This could have two very important consequences. The first consequence is that gaps where expert systems do not currently exist or where existing expert systems are demonstrated to be inadequate in scope or robustness, can be identified, located, or plotted in a systematic and planned manner, in effect on a map. For complex phenomena this is a fundamental exercise (e.g. radio astronomers have gradually built up a complex map of the observed universe over many years). Much of any initial work concerned with the development of the algorithm for a new or improved expert system will have already been completed as part of the SDA process.

The second consequence is that the SDSM will demonstrate where interaction between any stand-alone expert systems might be fruitful.

Implications of SMSMs for robotics

Any modern cockpit flight management system on board an airliner needs to know, among other things, and on a continuous basis, the exact position of the aircraft, and this in three dimensions. The implications drawn from a computer-based assessment of these data will affect the assessment of other or future data generated or automatically captured. It will affect any decision on the importance of data such as fuel reserves, navigation heading corrections, and any other flight adjustments such as safe flying altitude. In just the same way, within the construction process a truly 'smart' robot, as opposed to a quasi-robot with umbilical cord to a remote control console, will need to know, among other things, its exact position on some kind of appropriate map, and will use this knowledge for the identification and a decision on the relevance, importance and priority afforded to other data, tasks and robots. Clearly if a truly smart robot, or a robot management system for a gang of smart robots, had access to an SDSM, and one that had been personalized for a specific project, it would open up a number of exciting possibilities for the robot's use.

For example, Hasegawa describes the development of a complete steel erection system which consists of a family of robots with a frame-climbing robotic 'platform'(ref. 66). The family of robots work in a co-ordinated way in positioning, adjusting and welding columns and beams. This is a step forward from the stand-alone robot, and likely to lead to more complex robot groups. A precedent for this is the personal computer, initially a stand-alone device which has been gradually integrated into large and complex networks that allow the user to benefit from both this and the stand-alone facility. The benefits of access to large computing facilities and the ability to achieve EDI make such a development very attractive.

Paulson *et al.* support this view in their discussion on the extent to which robots must be aware of knowledge about their environment, suggesting that there is a crucial level of knowledge that the robots must have to work on a project effectively (ref. 67). They also suggest that the ability to define such a knowledge base would enable managers to design and then to simulate data interchange between individual robots (or robot families) and their environment.

The BUILDING process and holistic design of management information systems

Most construction industries have evolved their systems and procedures over many centuries, and have thrown off surprisingly few of their medieval practices. Many ways of working can be shown by historians to have come about slowly and only in response to sustained client, technological, economic, national emergency, or market pressure.

By using SDA to develop a full SDSM of the total construction process, showing best current practice in each component GDFM, the systems engineer will have the opportunity to design or put together in a holistic way, from a series of specialist DFMs, an efficient, new total construction process that can make effective use of the benefits of modern electronic technology. This method of research was advocated in the *2001 series*, where identifying the best of current practice was deemed the appropriate way forward (ref. 6).

Design of integrated management EDP systems for computer-integrated construction

The construction design and fabrication process is complex and difficult to organize and manage for several reasons.

(a) The fragmented nature of the industry, and as a direct result the way in which it organizes itself, leads to poor design management, and concentration on the tender rather than the exit price. Also, bad communication and co-ordination between all parties to the project is often noticeable, usually because of a lack of holistic project leadership.
(b) The nature of the product, despite the impact that buildings have on society and the surrounding environment, means that project teams come together only for a specific period, and do not normally have a multi-project existence.
(c) The nature and relatively low level of craft skills of the people that work in the industry, and the lack of serious middle and senior management training across the industry compares unfavourably with other industries.
(d) The adversarial culture that too often exists at all levels within the industry means that ingenuity is directed at winning the 'inter-project team' arguments and pinning the blame and the risk, rather than on getting the project built and performing to the client's total satisfaction.
(e) Each project is largely unique, with key parts of it built in the open air, and built on a different geographical site.
(f) The bad image the industry has with clients (regular and irregular), opinion formers, the general public and the new generation has possibly resulted in a low level of self-worth among those employed in the industry, and a low quality of recruits attracted to it.

This catalogue of factors has affected the performance of the industry in one way or another, and has hindered attempts to change and improve it over many years. Clients, particularly those in the private sector, are too often dissatisfied with the performance that they receive from the industry (ref. 6). The major Japanese and possibly French contractors are responding by moving rapidly towards computer-integrated construction. Around 80% of their work is based on a direct relationship with their clients, involving some form of design, management and construction. A considerable amount of Japanese research effort, involving many of Japan's most talented engineers and architects, is directed at developing such systems. Much of the current work is in the areas of more flexible CAD, but practical applications have so far been limited. Despite this, their commitment is undeniable and significant progress has been made. With design and construction within the same large construction company, the task facing the Japanese is perhaps easier than that facing their EC counterparts.

Computer-integrated construction does, however, offer a medium to long-term solution to most of the difficulties outlined, and one that is capable of offering a considerable productivity advantage, which will harness in a cost-effective way the benefits of expert systems and robotics. As has been stressed, SDA can offer considerable benefit as a research tool in the complex task of developing real computer-integrated construction. This step forward, using SDA, DFMs, GDFMs, and an SDSM of the overall process, will enable the building process and the systems design of integrated management EDP systems for computer-integrated construction to be developed rapidly and in a systematic and effective way.

References

1. A Financial Times survey. Expanding into new areas. *Financial Times*, London, 5 June 1992, 1-4.
2. A Financial Times survey. Computers in manufacturing. *Financial Times*, London, 24 Sept. 1992, 31-32.
3. WOMACK, J. P. *et al. The machine that changed the world.* Rawson Associates, New York, 1990.
4. GARAS, F. K. Automation and robotics in construction - an overview of R&D in the UK. *Proc. 9th Int. Symp. Automation and Robotics in Construction, Tokyo*, 1992, 45-49.
5. NATIONAL ECONOMIC DEVELOPMENT OFFICE. *Information transfer in building.* NEDO, London, 1991.
6. UNIVERSITY OF READING CENTRE FOR STRATEGIC STUDIES IN CONSTRUCTION and THE NATIONAL CONTRACTORS 'GROUP. *2001 series.* CSSC, University of Reading and Building Employers' Confederation, includes unpublished Task Force report, 1990.
7. ABDUL RASHID AZIZ. *The globalization of construction.* PhD thesis, University of Reading, submitted 1991.
8. WALD, J. and TATUS, C. B. Diffusion of construction technology. *Proc. 9th Int. Symp. Automation and Robotics in Construction, Tokyo*, 1992, June, 73-82.
9. KOSKELA, L. Process improvement and automation in construction: opposing or complementing approaches? *Proc. 9th Int. Symp. Automation and Robotics in construction, Tokyo*, 1992, June, 105-112.
10. GARAS, F. K. (ed.) *Proc. 9th Into. Symp. Automation and Robotics in Construction, Tokyo*, 1992, June.
11. ASKEW, W. H. *et al.* Automating and integrating the management function. *Proc. 9th Int. Symp. Automation and Robotics in Construction, Tokyo*, 1992, June, 385-392.
12. FISHER, N. The use of structured data analysis as a construction management research tool: 1 - the technique. *Construction management & economics*, 1990, **8**, 4, 341-363.
13. FISHER, N. A structured analysis of data flow systems for client-based construction project management and for contracting companies. Thomas Telford, London, 1991, 63-76.
14. SEWARD, D. *et al.* Controlling an intelligent excavator for autonomous

digging in difficult ground. *Proc. 9th Int. Symp. Automation and Robotics in Construction, Tokyo*, 1992, June, 743-750.
15. *Proc. 8th Int. Symp. Automation and Robotics in Construction, Stuttgart*, 1991, Summaries, 31-32; 43-45.
16. SASAKI, Y. and KANAIWA, T. An integrated construction management system for site precast concrete. *Proc. 8th Int. Symp. Automation and Robotics in Construction, Stuttgart*, 1991, June 399-408.
17. SALAGNAC, J. -L. Construction robotics - state of the art in France. *Proc. 8th Int. Symp. Automation and Robotics in Construction, Stuttgart*, 1991, June 54-57.
18. JOLY, J. -M. Vers l'automatisation des operations de manutentions sur les chantiers de construction. *Colloque CAD et robotique en architecture et BTP, CSTB, GAMSAU, IIRIAM, Marseille, 1986*. Hermes, Paris, 1986.
19. *Proc. 8th Int. Symp. Automation and Robotics in Construction, Stuttgart*, 1991, Summary 30.
20. SHOHET, I. M. and LAUFER, A. Development of an expert system for the determination of the span of the control of the construction foremen. *Proc. 8th Int. Symp. Automation and Robotics in Construction, Stuttgart*, 1991, 507-517.
21. SKIBNIEWSKI, M. J. Current status of construction automation and robotics in the United States of America. *Proc. 9th Int. Symp. Automation and Robotics in Construction, Tokyo*, 1992, 17-24.
22. MOSELHI, O. *et al.* A hybrid neural network methodology for cost estimation. *Proc. 8th Int. Symp. Automation and Robotics in Construction, Stuttgart*, 1991, 519-528.
23. *The Economist*. 1992, 325, No. 7763, 13-19 June, 148-149.
24. KARLEN, I. *Information methodology - information management*. Information Study Group, CIB Publications 65, Stockholm, 1982, 134-175.
25. Boston Consulting Group. Globalisation of construction. Unpublished study, Paris, 1990.
26. FISHER, N. and SHEN LI YIN. The use of structured data analysis as a construction management research tool: 2 - Field trials. *Construction management & economics*, submitted for publication.
27. MORRIS, P. W. G. *The anatomy of major projects*. Wiley, Chichester, 1988.
28. CHARTERED INSTITUTE OF BUILDING. *The Code of estimating practice*. CIOB, Ascot, 1965.
29. CHORLEY, R. J. and KENNEDY, B. A. *Physical geography: a system approach*. Prentice-Hall, London, 9171.
30. CHAPMAN, G. P. *Human and environmental systems*. Academic, New York, 1977.
31. BENNETT, R. J. and CHORLEY, R. J. *Environmental system*. Methuen, London. 1978.

32. BOLTON, D. *et al. Systems behaviour - modules.* Open University, Milton Keynes, 1977, 8-87.
33. NAUGHTON, J. and PETERS, G. Systems and failures - human factors & systems failures. *Technology : a third level course, Unit 1.* Open University, Milton Keynes, 1976, 19-27.
34. MEDAWAR, P. B. *Advice to young scientists.* Harper & Row, London, 1979.
35. POPPER, K. R. *Conjectures and refutations: the growth of knowledge.* Routledge & Kegan Paul, London, 1963.
36. POPPER, K. R. *Objective knowledge.* Oxford University Press, 1972.
37. KUHN, T. *The structure of scientific revolutions,* 2nd edn. Chicago University Press, 1970.
38. YOURDON, E. *Structured walkthroughs.* Yourdon, New York, 1979.
39. WALKER, A. and HUGHES, W. P. A Project managed by a multidisciplinary practice: a system-based case study. *Construction management & economics,* 1987, **5**, 123-140.
40. MUNDAY, M. *Education for information management in the construction industry.* HMSO, London, 1978, Report of a study by the University of Strathclyde for the Department of the Environment and the National Consultative Council of the Building and Civil Engineering Industries Standing Committee on Computing and Data Co-ordination.
41. MUNDAY, M. User education - an introduction to the construction information education project. *Proceedings of conference on user education research,* Loughborough University, 1979.
42. BALL, M. *The contracting system in the construction industry.* Discussion paper 86, Birkbeck College, University of London, 1980.
43. BOWLEY, M. *The British building industry.* Oxford University Press, 1963.
44. BISHOP, D. Building technology in the 1980s. *Phil. Trans. R. Soc.*, London, **A272** 533-563.
45. PORTER, L. W. *et al. Behaviour in organizations* McGraw-Hill Kogakusha, Tokyo, 1975.
46. CHECKLAND, P. B. *Systems thinking systems practice.* Wiley, Chichester, 1981, 96-97.
47. JACKSON, M. A. *Principles of program design.* Academic, London, 1975.
48. DE MARCO, T. *Structured analysis and system specification,* 2nd edn. Yordon, New York, 1979.
49. DAHL, O. J. *et al. Structured programming.* Academic, New York, 1972.
50. MYERS, G. J. *Reliable software through composite design.* Petrocelli/Charter, New York, 1976.
51. YOURDON, E. and CONSTANTINE, L. L. *Structured design.* Yourdon, New York, 1978.
52. STEVENS, W. P. *et al.* Structured design. *IBM Systems J.* 1974, **13**, No. 2, 115-139.
53. FISHER, G. N. Towards a general model of project monitoring and control

systems as used by broadly similar and successful building contractors. *Proc. CIB W65*, 1984, **4**, July, Waterloo, Canada.

54. FISHER, G. N. *Project monitoring and control systems as used by broadly similar and successful contractors.* Science and Engineering Research Council, Swindon, 1985.
55. FISHER, G. N. *Marketing for the construction industry - a handbook for consultants, contractors and other professionals.* Longman, London, 1989.
56. FISHER, G. N. and ATKIN, B. L. Computer workstation for construction sites. *Int. J. Construction Mgmt & Technol.*, 1986, **1**, No. 2, 17-29.
57. FISHER, G. N. The use of 'Structured Data Analysis' as a design tool for computer integrated construction. *Proc. CIB W74 and W78 Congr. on Computer Integrated Construction, Tokyo*, 1990, 263-269.
58. DICKINSON, B. *Developing structured systems.* Yourdon, New York, 1980.
59. FLAVIN, M. *Fundamental concepts of information modelling.* Yourdon, New York, 1981, 21-109.
60. WARDM P. T. *Systems development without pain.* Yourdon, New York, 1984, 14-75.
61. KEFALASAND, L. and SCHODERBEK, C. *The elements of programming style.* Bell Telephone Laboratories, Murry Hill, New Jersey, 1974.
62. KELLER, R. *The practice of structured analysis.* Yourdon, New York, 1983.
63. BOHM, C. and JACOPINI, G. Flow diagrams, tutor machines, and languages with only two formation rules. *Communications of the ACM.* The National Science Foundation, Washington D.C., 1966, May, 366-371.
64. MCMENAMIN, S. M. and PALMER, J. F. *Essential systems analysis.* Yourdon, New York, 1984, 49-68, 204-226.
65. OPEN UNIVERSITY SYSTEMS GROUP. Papers 1981-9. Open University, Milton Keynes.
66. HASEGAWA, Y. *et al.* Development of a new steel erection system. *Proc. 6th Int. Symp. Robotics and Automation in Construction, San Francisco, 1989.* Construction Industry Institute, San Francisco, 515-522.
67. PAULSON, B. *et al.* Simulating the knowledge environment for autonomous construction robot agents. *Proc. 6th Int. Symp. Robotics and Automation in Construction, San Francisco, 1989.* Construction Industry Institute, San Francisco, 475-482.

Glossary

Balancing The relationship that exists between parent and child diagrams in a properly levelled DFD set; specifically the equivalence of input and output data flows portrayed at a given process on the parent diagram and the net input and output data flows on the associated child diagrams.

Context boundary An arbitrary boundary set by a systems engineer between two or more DFMs and in particular around a DFM that is the subject of study or partitioning. All transmitters or receivers of information that are not enclosed within the context boundary are terminators as far as that DFM is concerned.

Context diagram Top level diagram of a levelled DFD set; a DFD that portrays all the net inputs and outputs of a system, but shows no partitioning.

Data dictionary (DD) A set of definitions of data flows, files, information sources/sinks/terminators, and processes referred to in the levelled DFD set.

Data flow A pipeline along which information of known composition is passed.

Data flow diagram (DFD) A network of related functions showing all interfaces between components; a partitioning of a system and component parts.

Data flow model (DFM) A representation of a specific company system using DFDs and a DD.

Electronic Data Interchange (EDI)

File A data store, repository of data; a time-delayed data flow

Functional Primitive Lowest level component of a Data Flow Diagram, a process that cannot be further decomposed to a lower level.

General Data Flow Model (GDFM) The combination of two or more DFMs drawn from fieldwork, in, for example, a business organisation, into one common system. A GDFM represents the best of current practice in two or more observed business organisations. The most complete or comprehensive of the observed sub-systems within the individual DFMs are combined into one GDFM.

Interface points Interface points exist at points of information exchange between different systems or sub-systems, usually DFMs, GDFMs or SDSMs. This information is exchanged across a context boundary between discrete sub-systems.

Information Technology (IT)

Mini specification (Mini specs) A statement of the policy governing the transformation of input data-flow(s) at a given functional primitive.

Process The transformation of input data flow(s) into output data flows(s).

Sink A net receiver of information outside of the subject system.

Source A data source outside of the subject system.

Structured Data System Map (SDSM) A combination of two or more GDFMs to form a map of the data flows of part of a larger more complex system, or a whole system such as the construction procurement process.

Terminator A net receiver or supplier of information outside of the subject system (i.e. a source or a sink).

Acknowledgements

The authors thank the following for their invaluable contribution and/or support, often over a very long period.

For financial support: the Science and Engineering Research Council (SERC), the Chartered Institute of Building (CIOB) QE2 scholarship and the University of Reading.

For invaluable help and advice at the analysis, synthesis and development stages: Professor John Bennett DSc; numerous colleagues in our own department of construction management & engineering and others at the University of Reading in the departments of applied statistics, computer science, economics (including the finance and accountancy section) and geography; those closely involved from Hewlett-Packard Ltd, CAP (Systems Group) Ltd, Ford (UK) Ltd, Yourdon (UK) Ltd, members of the Yourdon User Group and the companies that took part in the study but wished to retain their anonymity. Also Professor Roger Burgess at the University of Liverpool, and Mike Ankers and Peter Harlow at the CIOB for their help and encouragement.

At the early fieldwork, raw data collection and analysis stages: N.C. Tinker, BSc, MSc; numerous members of the MSc course in project management; doctoral students and a number of undergraduates in particular on the honours degree courses in building construction and management and quantity surveying. For typing the manuscript: Jean Ninian.

About the authors

Dr Norman Fisher is the BAA Professor of Construction Project Management in the Department of Construction Management & Engineering at the University of Reading, UK, and is a Chartered Builder. He has wide industrial and research experience. He has undertaken extensive marketing, business consultancy, and related fundamental and applied research. He lectures regularly to university and senior industry audiences worldwide.

Dr Shen Li Yin obtained his BSc in Construction Engineering at the Chongqing Institute of Architecture and Engineering, Chongqing, China, where he is a lecturer. He worked as a post-doctoral research fellow in the Department of Construction Management & Engineering, University of Reading, UK, during 1990 and 1991. He is now on secondment to the University of Hong Kong.

CPSIA information can be obtained
at www.ICGtesting.com
Printed in the USA
BVHW041424100620
581228BV00006B/92

9 780727 716668